STATISTICAL DETECTION AND SURVEILLANCE OF GEOGRAPHIC CLUSTERS

T0225647

CHAPMAN & HALL/CRC
Interdisciplinary Statistics Series

Series editors: N. Keiding, B.J.T. Morgan, C.K. Wikle, P. van der Heijden

Published titles

AN INVARIANT APPROACH TO STATISTICAL ANALYSIS OF SHAPES	S. Lele and J. Richtsmeier
ASTROSTATISTICS	G. Babu and E. Feigelson
BAYESIAN DISEASE MAPPING: HIERARCHICAL MODELING IN SPATIAL EPIDEMIOLOGY	Andrew B. Lawson
BIOEQUIVALENCE AND STATISTICS IN CLINICAL PHARMACOLOGY	S. Patterson and B. Jones
CLINICAL TRIALS IN ONCOLOGY SECOND EDITION	J. Crowley, S. Green, and J. Benedetti
CORRESPONDENCE ANALYSIS IN PRACTICE, SECOND EDITION	M. Greenacre
DESIGN AND ANALYSIS OF QUALITY OF LIFE STUDIES IN CLINICAL TRIALS	D.L. Fairclough
DYNAMICAL SEARCH	L. Pronzato, H. Wynn, and A. Zhigljavsky
GENERALIZED LATENT VARIABLE MODELING: MULTILEVEL, LONGITUDINAL, AND STRUCTURAL EQUATION MODELS	A. Skrondal and S. Rabe-Hesketh
GRAPHICAL ANALYSIS OF MULTI-RESPONSE DATA	K. Basford and J. Tukey
INTRODUCTION TO COMPUTATIONAL BIOLOGY: MAPS, SEQUENCES, AND GENOMES	M. Waterman
MARKOV CHAIN MONTE CARLO IN PRACTICE	W. Gilks, S. Richardson, and D. Spiegelhalter
MEASUREMENT ERROR AND MISCLASSIFICATION IN STATISTICS AND EPIDEMIOLOGY: IMPACTS AND BAYESIAN ADJUSTMENTS	P. Gustafson

Published titles

Chapman & Hall/CRC
Interdisciplinary Statistics Series

STATISTICAL DETECTION AND SURVEILLANCE OF GEOGRAPHIC CLUSTERS

Peter Rogerson

University at Buffalo
Buffalo, New York, U.S.A.

Ikuho Yamada

University of Utah
Salt Lake City, Utah, U.S.A.

CRC Press
Taylor & Francis Group
Boca Raton   London   New York

CRC Press is an imprint of the
Taylor & Francis Group, an **informa** business

A CHAPMAN & HALL BOOK

Chapman & Hall/CRC
Taylor & Francis Group
6000 Broken Sound Parkway NW, Suite 300
Boca Raton, FL 33487-2742

First issued in paperback 2020

ISBN-13: 978-0-367-57741-4 (pbk)
ISBN-13: 978-1-58488-935-9 (hbk)

Visit the Taylor & Francis Web site at
http://www.taylorandfrancis.com

and the CRC Press Web site at
http://www.crcpress.com

Contents

List of Figures

List of Tables

Acknowledgments

In writing this book, we have had the benefit of both helpful conversations and able research assistance. Over the years in which the ideas contained in this book were in their formative stages, many people have contributed in these and other ways. We would like to thank, in particular, Martin Kulldorff, Andrew Lawson, Marianne Frisen, Tom Talbot, Gwen Babcock, Ken Kleinman, Daikwon Han, Gyoungju Lee, Gaurav Sinha, and Yifei Sun. Both David Grubbs and Amy Blalock at CRC Press were helpful and a pleasure to work with. It is difficult if not impossible to make lists such as these complete, and we acknowledge our indebtedness to others who have contributed in various ways.

Much of the work described in this book has been facilitated by the helpful financial support from a number of organizations. The support of a Guggenheim Fellowship to the first author is gratefully acknowledged. Financial assistance was also provided through National Science Foundation Grant SBR88-10917 to the National Center for Geographic Information and Analysis. The support received from National Cancer Institute Grant R01 CA92693-01 is gratefully acknowledged, as is the support of Award 98-IJ-CX-K008 from the National Institute of Justice and the cooperation of the Buffalo Police Department. Some of the material is based upon work supported by the National Science Foundation under Award No. BCS-9905900, and some is also based upon work supported under grant number 1 R01 ES09816-01 from the National Institute of Environmental Health Sciences, National Institutes of Health. The second author acknowledges both a grant from the Southwestern Consortium for Environmental Research and Policy, and the Center for Health Statistics, California Department of Health Services for providing data used in one of the examples.

A significant portion of this book is based upon the culmination of work carried out over the last decade. Publishers were contacted, and they have granted permission for material from previous papers and book chapters to be used here. The correspondence between sections of this book, and the papers and book chapters some of them are based upon, is as follows:

Section	Source:
1.3.1	Rogerson (2006c)
2.2–2.5	Rogerson (2006c; Sections 2.6.1–2.6.4; pp. 32–35)
Parts of 3.4	Rogerson (2006c; Section 10.3.2; pp. 232–236)
3.11	Rogerson (2006a)
3.12	Rogerson (2006a)

4.8	Rogerson (2006a)
5.3	Rogerson (2004)
5.10	Rogerson (2001a)
6.4.2	Han and Rogerson (2003)
6.4.3	Rogerson, Sinha, and Han (2006)
7.4	Rogerson and Yamada (2004b)
7.8	Rogerson (2006b)
8.3	Lee and Rogerson (2007)
8.4.1	Rogerson (1997)
8.4.2	Rogerson (2001b)
8.4.3	Rogerson and Sun (2001)
9.1.3	Rogerson (2009a)
9.1.4	Rogerson (2000)
9.1.5	Rogerson (1997)
9.1.7	Rogerson (2009b)
9.2	Rogerson (2005c)
10.2	Rogerson (2005b)
10.3	Rogerson and Yamada (2004a)
Chapter 11	Rogerson (2005a)

In addition, the introductory material in Chapter 7, some of the material in Section 7.5 on the Poisson distribution, Section 7.6 on the exponential cusum, and Section 7.7 on other modifications of cusum charts are based upon Rogerson (2005c).

1

Introduction and Overview

1.1 Setting the Stage

Imagine yourself in the role of a public health official in an urban area. You are asked to monitor data on admissions to emergency departments for respiratory distress and to report on any abnormal increases that may occur. As part of your response, you decide to both tabulate and map the daily, weekly, and monthly reports of such occurrences. As you examine your first map depicting the location of new cases, how do you decide whether the map of occurrences has an interesting pattern? There may be apparent clusters of incidents, but these may simply reflect geographic patterns of population density. Perhaps you have accounted for population density and also for the age structure of the population. Is there still a pattern on the map that deviates significantly from some simple random assignment of cases to the population? The geographic pattern may reflect other factors—perhaps individuals living near hospitals are more likely to be admitted for respiratory distress.

Once issues associated with the detection of geographic pattern on any specific map are sorted out, other issues then arise in the repeated analysis of such maps. How different is today's map from yesterday's, and from the map that was "expected" based on past information? How do we decide whether daily, weekly, or monthly fluctuations in the geographic pattern are the result of simple random variation or the consequence of some more noteworthy event (such as a rapid decline in air quality in a particular part of the region)?

Researchers in other fields face similar questions. Crime analysts have exchanged their well-worn wall maps containing colored pins indicating the locations of crimes for new geographic information systems (GIS) that store information on crime locations electronically. However, the questions remain the same—is there a pattern to crime locations? Often there is—there are more crimes in particular types of neighborhoods, for example—and the more interesting question is whether there are geographic patterns in the crime location data once we have accounted for known influencing factors such as type of neighborhood. It would also be useful to know how we might monitor daily data on, for example, burglaries or arsons, so that we may decide when geographic patterns have changed; this could then lead to a reallocation of enforcement effort.

Market researchers are interested in delineating the size and location of the market for their products. They are also interested in how market areas change over time as the strength of their competitors waxes and wanes. For example, an analyst for a supermarket chain may plot the location of customers on a map and may monitor the locations of new customers (as well as those of customers they are losing). Are the locations of new customers different from those of existing customers? Where is the market getting stronger or weaker, and why?

All of these questions demand both data and statistical analysis. Although visualization is an important approach to exploring geographic data, it alone is not sufficient to answer the kinds of questions just suggested. In fact, visualization alone can actually be misleading; people tend to see clusters and patterns on maps where none exist, and they tend to overestimate the significance of patterns and clusters that do exist.

The purpose of this book is to provide statistical tools for (1) deciding whether the data on a given map deviate significantly from expectations (e.g., based on the location and characteristics of the population), and (2) deciding quickly whether new map patterns are emerging over time. Many previous efforts have been devoted to the statistical analysis of patterns on a single map, and so the content related to the first-stated purpose will constitute a review of selected, established methods. There has been much less work on the problem of monitoring geographic data over time, where the objective is the quick detection of spatial change. This new field of spatial pattern analysis is called *spatial surveillance* (or *spatial monitoring*). There has been a recent surge in interest in these kinds of questions, and we will spend time reviewing both basic approaches and recent developments in the field of spatial surveillance.

1.2   The Roles of Spatial Statistics in Public Health and Other Fields

For many fields, an understanding of spatial statistics is important for at least two general reasons. First, standard (nonspatial) statistical approaches must be modified when applied to spatial data due to the lack of independence in such data. Simply put, observations located near one another in space are often highly correlated. They are not independent; instead, they exhibit spatial dependence. Such commonplace methods as difference-of-means tests, correlation, and regression—when applied to spatial data—can often lead to misinterpretation because of their reliance on the assumption of independent observations, which is often violated in spatial data. Thus, if we have n responses to a questionnaire, we may not have n *effectively independent* observations; two individual respondents who live near to each other are likely to give similar answers. This positive spatial dependence leads us to have, effectively, fewer than n independent pieces of information. Consequently, proceeding with statistical tests in the presence of spatial dependence may cause us to reject null hypotheses when they are true; the rejection occurs due merely to the spatial dependence and not to deviations from the null hypothesis. Alternatively stated, if we proceed as if we have more information (n independent observations) than we really do (effectively, fewer than n independent pieces of information), our confidence intervals for mean differences, correlation, and regression coefficients, etc., will be narrower than they should be, occasionally causing us to wrongly reject true null hypotheses and to uncover seemingly significant relationships that should instead be attributed to the spatial dependence. An important role of spatial statistics, then, is to generalize standard, nonspatial statistical methods to address the lack of independence, and to account for spatial dependence.

A second use of spatial statistics is to assist in the search for spatial patterns in geographic data. In particular, it is often of interest to decide whether clusters of events exist on maps—examples include deciding whether there are significant geographic clusters of disease, crime, or customers for a store. Spatial statistical methods have been developed to test the null hypothesis of no spatial clustering. It is this latter use of spatial statistics to test hypotheses related to spatial clustering that constitutes the primary focus of this book.

1.3 Limitations Associated with the Visualization of Spatial Data

1.3.1 Visual Assessment of Clustering Tendency

In this section, we will see (1) that the eye is not particularly good at differentiating statistically significant spatial clusters from random spatial patterns, and (2) that there are different ways to map data, and that these alternative visual depictions can give rise to very different interpretations.

The following exercise is taken from Rogerson (2006c):

> Draw a rectangle that is six inches by five inches on a sheet of paper. Locate 30 dots at random within the rectangle. There are two characteristics of randomness: (1) for each point you locate, the likelihood that a subregion receives a point should be proportional to the size of that subregion, and (2) each dot should be located independently of the other dots.
>
> Next, draw a six-by-five grid of 30 square cells on top of your rectangle. You can do this by making little tick marks at one-inch

intervals along the sides of your rectangle. Connecting the tick marks will divide your original rectangle into 30 squares, each having a side of length one inch.

Give your results a score as follows. Each cell containing no dots receives one point. Each cell containing one dot receives 0 points. Each cell containing two dots receives 1 point. Cells containing three dots receive 4 points, cells containing four dots receive 9 points, cells containing 5 dots receive 16 points, cells containing 6 dots receive 25 points, and cells containing 7 dots receive 36 points. Find your total score by adding up the points you have received in all thirty cells.

On average, a set of 30 randomly placed points will receive a score of 29. A set of randomly placed points will receive, 95% of the time, a score between 17 and 45. The majority of people who try this experiment produce patterns that are more uniform or regular than random, and hence their scores are less than 30. Their point patterns are more spread out than a truly random pattern. When individuals see an empty space on their diagram, there is an almost overwhelming urge to fill it in by placing a dot there! Consequently, the locations of dots placed on a map by individuals are not independent of the locations of previous dots, and hence an assumption of randomness is violated.

Consider next Figure 1.1a and 1.1b, and suppose you are a health official looking at the spatial distribution of recent disease. Indicate "in pencil where you think the clusters of disease are. Do this by simply encircling the clusters (you may define more than one on each diagram)." Assume that the background population is spatially uniform.

How many clusters did you find? It turns out that both diagrams were generated by locating points at random within the rectangle! In addition to having trouble drawing random patterns, individuals also have a tendency to 'see' clusters where none exist. This results from the mind's strong desire to organize spatial information.

The widespread popularity and use of GIS has led to many new insights in countless areas of application. GIS has facilitated not only the collection and storage of geographic data but also the display of such data. Visualization is an important early phase of understanding the nature of spatial data. However, to be used effectively, GIS must be used for both the display and analysis of data. These exercises point to the need for objective, quantitative measures of spatial pattern—it is simply not sufficient to rely on one's visual interpretation of a map. Crime analysts cannot necessarily pick out true clusters of crime just by looking at a map, nor can health officials always pick out significant clusters of disease from map inspection. Such visual investigation clearly becomes much more difficult when the underlying population is not uniformly distributed over the study region.

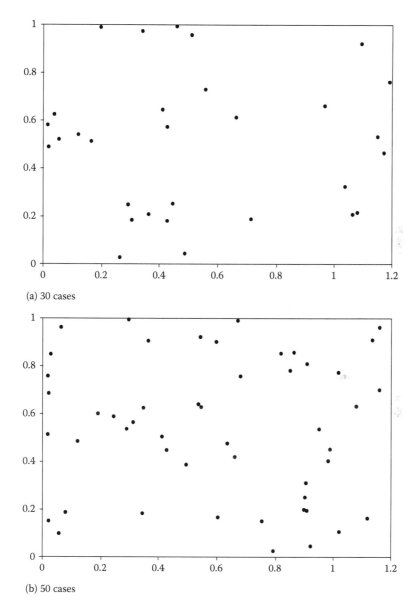

FIGURE 1.1

Hypothetical spatial distributions of disease cases: (a) 30 cases; (b) 50 cases.

1.3.2 What to Map: Mapping Rates versus Mapping *p*-Values

The initial choice of *what* to map is an important one because different statistical measures can lead to very different visual depictions, and hence, different interpretations.

It is not uncommon to map rates (e.g., of disease). These rates can be calculated in various ways; for example, as the ratio of the number of cases to the number of individuals at risk of being cases, or as the ratios of observed to expected cases (e.g., as in the standardized mortality ratio [SMR]). In both of these situations, extremes—either high rates or low rates—are more likely to occur where the expected number of cases is small. Thus, maps of rates and SMRs can often yield a perception of higher-than-average values when this is due to nothing more than the random variation associated with the low expectations that characterize these regions. Similarly, areas with a high expected number of cases will tend to look "average" on such maps. Even when the actual relative risk is somewhat raised in these areas, it is unusual to find the high rates or SMRs that can be found by chance in smaller regions.

An alternative (and less common) option is to map p-values (Choynowski 1959), which are defined as the probability of an observation that is equal to or more extreme than what is observed if the null hypothesis of no raised incidence is true. This has the advantage of accounting for the size of the denominator, and what is being mapped is a statistical measure of the unusualness of the observation, presuming no raised incidence. Maps of p-values are not without their drawbacks either. It is easier to attain statistical significance with large sample sizes, and maps of p-values, therefore, often mimic to some degree maps of population or expected numbers of cases. Specifically, there can be a tendency for areas of large population to have small p-values. This can occur when there are small departures from the underlying assumption of a Poisson model (e.g., see Cressie 1993).

1.3.2.1 Example 1: Sudden Infant Death Syndrome in North Carolina

Table 1.1 shows the rates of sudden infant death syndrome (SIDS) for selected counties of North Carolina in 1974. The statewide rate was 2.02 cases per 1000 births. The selected counties are arranged in the table

TABLE 1.1

Data on Sudden Infant Death Syndrome (SIDS) for Selected Counties in North Carolina, 1974

County	SIDS Rate (Per 1000 Births)	Probability (Significance) (p-Value)	Births
4	9.55	.0003	1570
66	6.33	.011	1421
94	5.05	.032	990
9	4.49	.029	1782
24	4.48	.005	3350
93	4.13	.103	968

Note: Statewide SIDS Rate in North Carolina: 2.02/1000 births.

according to SIDS rates, beginning with the highest. The *p*-values represent the probability of observing an equal or higher rate than what was actually observed, assuming that the statewide rate prevails in each county. Note that the rate of 4.48 cases per 1000 births, occurring in county 24, is based on a large number (3350) of births. A simple test of the hypothesis that the statewide rate holds in this county (based on Poisson probabilities) is rejected strongly ($p = 0.005$). The strength of the statistical significance is greater than that for counties 66, 94, and 9, despite the fact that SIDS rates are higher in these counties (6.33, 5.05, and 4.49 cases per 1000 births for the three respective counties).

1.3.2.2 *Example 2: Breast Cancer in the Northeastern United States*

From the compressed mortality data sets (see Section 1.7 for a more complete description), we identified 306,953 deaths from breast cancer in the northeastern United States, implying an annual average age-adjusted death rate of 31.87 per 100,000 females during the time period of 1968 to 1998. The average standard mortality ratio in the Northeast relative to the United States for the entire time period is 1.13, implying a mortality rate in the Northeast that is higher than the nationwide rate. Figure 1.2a is a map of the standardized mortality ratio using Equation 1.1, and Figure 1.2b shows a map of the *p*-values resulting from individual Poisson tests of randomness in each county using Equation 1.2:

$$SMR_i = \frac{x_i}{\lambda_i} \tag{1.1}$$

$$p(x_i) = \Pr(X_i \geq x_i) = \sum_{X_i = x_i}^{\infty} \frac{e^{-\lambda_i} \lambda_i^{x_i}}{x_i!}, \tag{1.2}$$

where the SMR in region *i* is the ratio of the observed number of deaths x_i to the age-standardized expected number of deaths λ_i in that region. The Poisson probability $p(x_i)$ represents the upper tail of the Poisson distribution with parameter λ_i.

The highest quantiles in Figure 1.2a are the areas showing a number of deaths from breast cancer that is more than 14% above the expected number. Figure 1.2b shows that many counties in the Northeast have an observed number of cases that is statistically significant relative to expectations, especially the darkest areas around New York and Boston. However, interpretation of these figures should proceed with caution. The SMR map (Figure 1.2a) is likely to have extreme values in areas with small populations. For example, the Adirondack region of New York State appears dark on the map in (a) but not in (b); its high SMR is not statistically significant. The map of *p*-values (Figure 1.2b) can also potentially be misleading

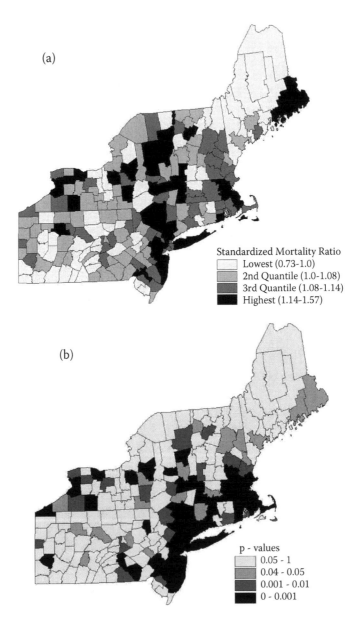

FIGURE 1.2
Mapping spatial variations of breast cancer mortality in the Northeast by county during 1968–1998: (a) standardized mortality ratio in the Northeast by county during 1968–1998; (b) Probability map (*p*-values).

because regions with large numbers of observed and expected cases are more likely to appear as significant (i.e., as dark areas) on such maps.

The point of this section is that one should exercise caution when interpreting maps visually. The eye tends to create clusters (to "organize" the spatial information) even when data are spatially random. In addition, maps of rates and maps of *p*-values can present very different visual depictions of spatial variability. Neither is "correct," and indeed some combination of the two is desirable. A high regional rate may not necessarily imply a low *p*-value; if the region's population is small, there will be greater uncertainty in the estimated rate. Maps presented in the *Atlas of Cancer Mortality* (Devesa et al. 1999), for example, show rates of disease, but provision is also made for conveying some measure of confidence or uncertainty in the rate.

1.4   Some Fundamental Concepts and Distinctions

1.4.1   Descriptive versus Inferential, and Exploratory versus Confirmatory, Spatial Statistics

Statistical analysis methods can generally be divided into two categories on the basis of their objectives: descriptive statistics and inferential statistics. The field of descriptive statistics aims to summarize the characteristics of a sample through the use of a small number of numerical indices and visualization methods. Nonspatial examples of descriptive statistics include the mean, median, and standard deviation, and frequency tables and histograms, to name only a few. The purpose of inferential statistics, on the other hand, is to make inferences about a population from characteristics of the sample. Statistical estimation, the construction of confidence intervals, and hypothesis testing belong to this category. Chapter 2 is devoted to descriptive and inferential statistics utilized in the context of spatial analysis.

A similar categorization of statistical methods is exploratory versus confirmatory. Exploratory methods are used to suggest hypotheses worth testing. For instance, mapping locations of cancer cases together with the locations of nuclear power plants may suggest a possible link between them, which in turn may help us form a hypothesis that the probability of getting cancer at a particular location is inversely proportional to the distance to the closest nuclear power plant. Confirmatory methods are used to evaluate and test given hypotheses. Exploratory methods often make use of descriptive statistics whereas confirmatory methods often make use of inferential statistics.

Although much emphasis has traditionally been placed on confirmatory analysis, there is also an increasing awareness of its limitations. Problems of interest are often too complex for confirmatory methods to simply confirm or refute their associated hypotheses. At the same time, the use of exploratory methods has increased rapidly as large and complex data sets as well as advanced analytical software including GIS has become widely available. This book discusses both exploratory and confirmatory methods with more emphasis on the latter.

1.4.2 Types of Health Data

1.4.2.1 Point Data

Locations of spatial entities (e.g., hospitals and disease cases) are often represented as points in a two-dimensional map space (e.g., Figure 1.1); such data are called *point data*. Each record in point data must have its positional information represented by the x- and y-coordinates, and may also contain additional attributes (for example, the square footage of the hospital, the number of physicians at each hospital, or the age, gender, and exposure level of each case). Disease cases represented by points may often contain intentional random errors in the case locations in order to protect the confidentiality of individuals.

1.4.2.2 Case–Control Data

We will also be interested in cluster detection and geographic surveillance when data are in the form of point locations for cases and controls. Here, cases refer to individuals with a particular disease of interest, and controls refer to individuals who have similar characteristics as cases (e.g., age, gender, and ethnicity) but do not have the disease. When controls can be seen as a representative subset of population without the disease, comparison of the spatial distribution of cases with that of controls helps us identify spatial patterns in the case distribution that are beyond what is merely reflective of the spatial distribution of the population. An example of case–control data is presented in subsection 1.7.2.5.

1.4.2.3 Areal Data

This is perhaps the most commonly available form of spatial health data because exact locations of disease cases are often not publicly releasable. With areal data, a study region is divided into a set of nonoverlapping zones (such as census tracts and ZIP code zones), and each zone has associated attribute values such as the number of disease cases and population. Again, to protect the confidentiality of individuals, when the number of cases in a zone is very small, it may not be reported. Areal data may also contain other physical and socioeconomic attributes associated with each zone, such as elevation, a measure of air pollution level, ethnic composition (e.g., percentage of population of each ethnic group), and median household income. The data set used in Example 2 in subsection 1.3.2.2 and shown in Figure 1.2 is a typical example of areal data.

1.4.2.4 *Time-Subscripted Data*

Although less commonly available, we will also be interested in data that have time subscripts in addition to positional information. For instance, a point dataset that represents each patient as a point on his or her residential address with associated date of diagnosis and an areal dataset that contains the number of cases and population for each zone over multiple years belong to this category of data. Such data offer us opportunities to examine not only spatial patterns in the data distribution but also spatial–temporal patterns and temporal changes in spatial patterns. When investigating a disease outbreak, for example, one's objective is not merely to detect spatial clusters of disease cases but also to identify how the size, shape, or location of the clusters is changing over time. The point data in Figure 1.1 are time-subscripted data if each point in the data has an associated time stamp (e.g., date of diagnosis); the areal data in Figure 1.2 are so if mortality rates or probabilities are recorded separately for each year from 1968 to 1998.

1.5 Types of Tests for Clustering

Besag and Newell (1991) proposed a three-way classification of methods for detecting the presence of spatial clustering.

"General" tests are designed to provide a single statistic that can be used to assess the degree to which a map pattern deviates from the null hypothesis of spatial randomness. Although they provide an indication of whether any observed clustering is significant, these methods do not provide additional information on the size and location of clusters.

"Focused" tests are used when one is interested in knowing whether there is a cluster of events around a single or small number of prespecified foci. For example, we may wish to know whether disease clusters around an incinerator, or crime clusters around a set of liquor establishments.

The third type of clustering methods described by Besag and Newell are "tests for the detection of clustering." These are used in the common situation in which there is no a priori idea of where and how large the clusters may be; the methods are aimed at searching the data and uncovering the location and size of any possible clusters.

General tests are carried out with *global* statistics; a single (global) summary statistic characterizes any deviation from a random pattern. *Local* statistics are used to evaluate whether clustering occurs around particular foci, and hence, are employed for focused tests. The third category of tests—tests for the detection of clustering—includes methods in which many local tests are carried out simultaneously. Local statistics may therefore be used to detect clusters either when the location is prespecified (focused tests) or when there is no a priori idea of cluster location (tests for the detection of clusters).

When a global test finds no significant deviation from randomness, local tests may be useful in uncovering isolated hot spots of increased incidence. When a global test *does* indicate a significant degree of clustering, local statistics can be useful in deciding whether (1) the study area is relatively homogeneous in the sense that local statistics are quite similar throughout the area, or (2) there are local outliers that contribute to a significant global statistic. Anselin (1995) provides more detailed discussion on the interpretation of local tests.

1.6   Structure of the Book

The contents of this book can conceptually be structured into three parts. The first part, consisting of Chapters 2 through 5, provides a review of the basic and important statistical approaches to detecting patterns in geographic data. It includes separate chapters on each of the three types of clustering tests described in Section 1.5. The second part consists of just one chapter, Chapter 6, and its aim is to review methods that are appropriate for the retrospective detection of change in spatial patterns. The third part, consisting of Chapters 7 through 10, focuses on geographic surveillance, where there is interest in the quick detection of change as geographic data are monitored over time. The first chapter in this third part is devoted to temporal monitoring; it is essentially a review of methods used to detect change in temporal data. These methods have their roots in industrial process control where there is concern with knowing rapidly any deviation from the desired production process. These are the methods that are adapted and used in the following chapters of the third part, which focus on geographic surveillance—the monitoring of spatial data as it becomes available, with the objective of finding emergent spatial patterns as quickly as possible. Methods to be discussed in this part can be seen as "prospective" approaches, in contrast to the "retrospective" methods to be reviewed in the first and second parts. The last chapter, Chapter 11, aims to summarize a variety of approaches discussed in this book by presenting a way of integrating retrospective and prospective detection of spatial clustering in a simplified and unified manner.

The book emphasizes application in public health more than applications in other areas. In part, this reflects the degree to which interest has developed in various fields of application. It is important to realize that the applications introduced in this book constitute specific uses of the methods that are described; the methods are general and can be applied to problems in many fields.

1.7 Software Resources and Sample Data

1.7.1 Software Resources

The following list of useful software resources is clearly selective and contains only those that are available free of charge; there are many excellent packages for spatial analysis and spatial statistics, in addition to those on this list.

1.7.1.1 GeoSurveillance

GeoSurveillance is a stand-alone software package that combines spatial statistical techniques and GIS routines to perform tests for the detection and monitoring of spatial clustering in both retrospective and prospective manners. Methods implemented in this software include, but are not limited to, score and M statistics for retrospective analysis, and cumulative sum control chart methods for prospective analysis. These methods will be explained in the first and third parts of this book, respectively. *GeoSurveillance* was developed by the authors and their collaborator (Gyoungju Lee). We provide illustrations of its use in various chapters of this book. Information URL: http://www.acsu.buffalo.edu/~rogerson/geosurv.htm

1.7.1.2 GeoDa

GeoDa is a stand-alone software package that provides exploratory spatial data analysis techniques for areal data. The development of this software has been led by Dr. Luc Anselin, and its functionality ranges from basic mapping to spatial autocorrelation statistics (both global and local) and spatial regression (Anselin et al. 2006). Information URL:https://www.geoda.uiuc.edu/

1.7.1.3 R

The R system is an integrated suite of software that serves as a programming language and environment for statistical computing and graphics (Venables et al. 2007). It can be expanded with add-on packages, and a variety of packages are available for spatial analysis and spatial statistics (e.g., see Bivand 2006). Information URL: http://www.r-project.org/

1.7.1.4 SaTScan

SaTScan is a stand-alone software package that implements spatial, temporal, and space–time scan statistics in both retrospective and prospective manners (Kulldorff 2006). The scan statistic is introduced in Chapter 5. This software does not have mapping functionality but provides outputs that are easily imported into a GIS environment. Information URL: http://www.satscan.org/

1.7.1.5 Cancer Atlas Viewer

Cancer Atlas Viewer is free software for exploring and analyzing spatial and temporal patterns in data from the National Atlas of Cancer Mortality of the National Cancer Institute (BioMedware 2005). Although its functionality ranges from basic mapping and exploratory data analysis to global and local cluster detection, this software is a limited version of *Space-Time Intelligence System* by TerraSeer (http://www.terraseer.com/) and applicable only for the National Atlas of Cancer Mortality data. Information URL: http://www.biomedware.com/

1.7.1.6 CrimeStat

CrimeStat is a stand-alone software package for the spatial statistical analysis of crime incident locations developed by Ned Levine & Associates and the National Institute of Justice (Levine 2007). Its functionality includes, but is not limited to, spatial descriptive statistics, spatial and space–time point pattern analysis, journey to crime analysis, and crime travel demand modeling. Although its primary objective is to support crime-mapping efforts, the software can be applied to other spatial applications, and provides outputs importable into a GIS environment. Information URL: http://www.icpsr.umich.edu/CRIMESTAT/

1.7.2 Sample Data Sets

Throughout the book we will use the following data sets to illustrate various methods.

1.7.2.1 Breast Cancer Mortality in the Northeastern United States

The study area is composed of 217 counties in 9 states in the northeastern United States. A map showing the locations of the 217 counties is given in Figure 1.3.

Breast cancer mortality data are taken from the National Center for Health Statistics' Compressed Mortality File (CMF). The CMF data are based on mortality files that provide statistics on all deaths recorded in the United States. These data are available at the county level for individual years for the period 1968–1998, grouped by age, sex, race, and all causes of mortality. Data are publicly available for the period 1968–1988; for the period 1989–1998, special application and permission are required. The data on deaths from breast cancer (International Classification of Diseases-9, code 174) were extracted for the 217 counties in the northeastern United States.

We calculated the expected number of breast cancer deaths using the population estimates provided in the CMF. The population estimates are based on Bureau of the Census estimates of county resident population (National Center for Health Statistics). The expected number of

FIGURE 1.3
The 217 counties included in the northeastern U.S. cancer mortality data.

breast cancer deaths in region i, λ_i, is calculated using the indirect standardization method by multiplying national age-specific death rates and the county population in each age group. That is,

$$\lambda_i = \sum_j p_{ij} d_j \tag{1.3}$$

where p_{ij} denotes the population at risk in age group j in region i, and d_j denotes the U.S. mortality rate from breast cancer for age group j.

1.7.2.2 Prostate Cancer Mortality in the United States

Prostate cancer mortality based on death certificates was obtained from the CMF produced by the National Center for Health Statistics. The CMF data are available at the county level for individual years for the period 1968–1998, grouped by age, sex, race, and underlying causes of death. The number of prostate cancer deaths (ICD-9 codes, 185.0–185.9) was obtained for each county in the contiguous United States from 1968 through 1998, for white men aged 25 and over, by 10-year age groups (because of the difficulties associated with analyzing small numbers of nonwhites outside of the South, we confined the data set to prostate cancer frequencies for white males). We excluded 30 deaths with unknown age information.

We derived the expected number of prostate cancer deaths for white males using the population estimates based on Bureau of the Census estimates of midyear county population provided with the CMF. The expected number of prostate cancer deaths in each county was obtained using indirect standardization (see Equation 1.3).

The number of deaths due to prostate cancer among white males rose steadily and doubled during the period 1968–1994, with approximately 14,000 annual deaths at the beginning of this period, and slightly more than 28,000 by the end of the period. By 1998, the annual number of deaths had fallen slightly, to about 26,500.

1.7.2.3   Sudden Infant Death Syndrome (SIDS) in North Carolina

The data on SIDS covers the 100 counties of North Carolina; there were 667 SIDS deaths during the period 1974–1978. This corresponds to a rate of death of 2.02 per 1000 births. The dataset contains x-y coordinates for county centroids, the number of cases for 1974–1978, and the corresponding number of births for each of the 100 counties.

This dataset was published as Table 6.1 of Cressie (1993).

1.7.2.4   Leukemia in Central New York State

The dataset consists of the number of leukemia cases in census tracts in an eight-county region of central New York State (census tracts are subareas with approximately 4,000 people). This region includes the cities of Syracuse, Ithaca, and Binghamton. The data were collected for the period 1978–1982; there were 592 cases across the 281 census tracts. The total population in the study area is 1,057,673, implying an incidence rate of 5.6 cases per 10,000 people over the period of study. The data are used in the book by Waller and Gotway (2004) and are available at http://www.sph.emory.edu/~lwaller/ch9index.htm.

A more detailed dataset, at the unit of the census block group, is available on a diskette that accompanies the book chapter by Waller et al. (1994). They are also available via the Internet at http://lib.stat.cmu.edu/datasets/csb/. This dataset consists of 790 rows, one for each block group. Each row contains information on the location of the geographic centroid, population, and number of leukemia cases. The reader may wish to note that rows 113 and 116 refer to the same centroid location, and so there are actually 789 unique locations.

1.7.2.5   Leukemia and Lymphoma Case–Control Data in England

This dataset of childhood leukemia and lymphoma cases formed the basis of the illustration given by Cuzick and Edwards (1990). There are 67 cases and 141 controls, collected for the period 1974–1986. The study area

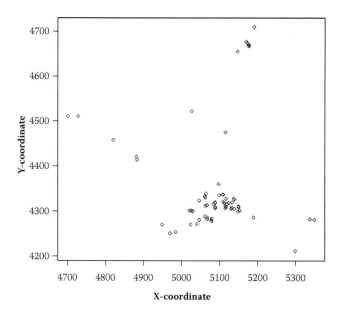

(a) Map of cases

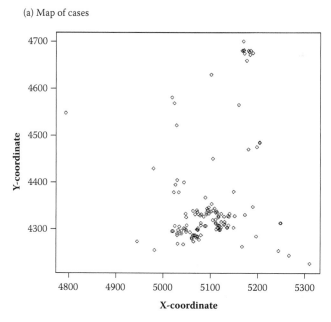

(b) Map of controls

FIGURE 1.4
Childhood leukemia and lymphoma cases and controls in the north Humberside region of England, 1974–1986: (a) map of cases; (b) map of controls.

is the north Humberside region of England. Controls were chosen from a sample of births from the birth register. Cases and controls are shown in Figure 1.4a and b, respectively.

1.7.2.6 *Low Birthweight in California*

This dataset is based on the California Birth Statistical Master File for the year 2000, obtained from the Center for Health Statistics, California Department of Health Services. The Master File contains all live birth records registered in California, and there are 532,845 records for 2000. An example dataset we will use in this book includes 482,333 successfully

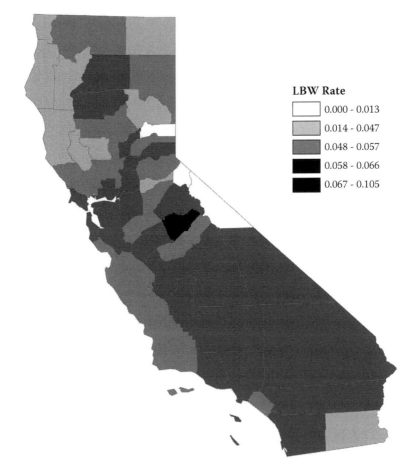

FIGURE 1.5
Low birthweight (LBW) rate in California, 2000, by county.

geocoded live births (approximately 90% of the total records) to mothers living in California. Among them, 29,635 cases have low birthweight (LBW, defined as less than 2,500 g), resulting in the LBW rate of about 6.1%, which is quite comparable with the U.S. rate of 6.5% in 2000 (CDC 2002).

Figure 1.5 shows the distribution of LBW rates aggregated into 58 counties in California.

2

Introductory Spatial Statistics: Description and Inference

2.1 Introduction

The study of (aspatial) statistics commonly begins with an introduction to *descriptive statistics*, which has as its aim the description of a set of data—typically using either visual or numerical methods. For instance, histograms provide visual summaries, and measures of central tendency (such as the mean and median) and variability (such as the standard deviation and variance) provide numerical summaries of a set of data.

We begin this chapter with a review of a number of descriptive statistics that are useful in providing numerical and visual summaries of *spatial* data. Descriptive measures of spatial data are important in assessing such fundamental geographic concepts as *accessibility* and *dispersion*. For example, it is important to locate public facilities so that they are accessible to defined populations. Spatial measures of centrality applied to the location of individuals in the population will result in geographic locations that are in some sense optimal with respect to accessibility to the facility. Similarly, it is important to characterize the dispersion of events around a point. It would be useful, for example, to summarize the spatial dispersion of individuals around a hazardous waste site. Are individuals with a particular disease less dispersed (and therefore more clustered) around the site than people without the disease? If so, this could indicate that there is increased risk of disease at locations near the site.

In this chapter, we also move beyond the realm of description into that of statistical inference. Although it is interesting to know where disease cases are clustered, it may also be of interest to test the hypothesis that they are clustered at a particular location specified in advance. While visualizing the spatial density of disease may be appealing, it is also important to test the hypothesis that the density surface does not exhibit more than random statistical fluctuations.

2.2 Mean Center

The most commonly used measure of central tendency for spatial data is the *mean center*, which is represented as a point location in a given study region. For point data, the x- and y-coordinates of the mean center are found by simply calculating the mean of the x-coordinates and the mean of the y-coordinates, respectively.

For areal data, one may still find a mean center for a set of representative locations of subregions (e.g., the centroids) while weighting each location according to an attribute of interest, such as population. To find the center of population, for instance, the weights are taken as the number of people living in each subregion. The weighted mean of the x- and y-coordinates then provides the location of the mean center. More specifically, when there are n subregions, or more generally n points, with associated weights w_i (e.g., population in region i),

$$\bar{x} = \frac{\sum_{i=1}^{n} w_i x_i}{\sum_{i}^{n} w_i} ; \quad \bar{y} = \frac{\sum_{i=1}^{n} w_i y_i}{\sum_{i}^{n} w_i} \tag{2.1}$$

where x_i and y_i are the coordinates of the centroid of region i or point i. Conceptually, this is identical to assuming that all individuals inhabiting a particular subregion live at a prespecified point (such as a centroid) in that subregion.

Each decade, the U.S. Bureau of the Census calculates the mean center of the population of the United States. Not surprisingly, the center has moved steadily west; it has also had a less pronounced yet noticeable southward drift (Figure 2.1).

The mean center has the property that it minimizes the sum of squared distances that individuals must travel to reach the mean center. Although

FIGURE 2.1

Historical mean centers of population in the United States, 1790–1990.

Note: The coordinates are found in U.S. Census Bureau (2001). Centroids for 1950 and later were calculated for coterminous United States only.

it is easy to calculate, this definition of centrality is a little unsatisfying; it is desirable to also be able to find a central location that minimizes the sum of distances rather than the sum of squared distances. Although the definitions of the two locations sound very similar, the locations are often different. For example, imagine a point distribution consisting of only two points. Any location along a straight line that connects the two points minimizes the sum of the distances, whereas only the middle point of the line minimizes the sum of the squared distances.

2.3 Median Center

The location that minimizes the sum of distances traveled is known as the *median center*. Although its interpretation is more straightforward than that of the mean center, its calculation is more complex. The calculation of the median center is iterative, and one begins by using an initial location (and a convenient starting location is the mean center). Then, the new x- and y-coordinates (denoted x' and y') are updated using the following:

$$x' = \frac{\sum_{i=1}^{n} \frac{w_i x_i}{d_i}}{\sum_{i=1}^{n} \frac{w_i}{d_i}} \; ; \quad y' = \frac{\sum_{i=1}^{n} \frac{w_i y_i}{d_i}}{\sum_{i=1}^{n} \frac{w_i}{d_i}} \tag{2.2}$$

where d_i is the distance from point i to the specified initial location of the median center. This same process is then carried out again—new x- and y-coordinates are again found using these same equations, with the only difference being that d_i is redefined as the distance from point i to the most recently calculated location (x', y') for the median center. This iterative process is terminated when the newly computed location of the median center does not differ significantly from the previously computed location.

In the application of social physics to human spatial interaction, the quantity, population divided by distance, is considered a measure of population "potential" or accessibility. If the w's are defined as populations, then note from Equation 2.2 that each iteration finds an updated location based on weighting each point or areal centroid by its accessibility to the current median center. The median center may therefore be interpreted as an accessibility-weighted mean center, where accessibility is defined in terms of the distances from each point or areal centroid to the median center.

2.4 Standard Distance

Aspatial measures of variability characterize the amount of dispersion of data points around the mean. Similarly, the spatial variability of locations around a fixed central location may be summarized. The *standard distance*

(Bachi 1963) is defined as the square root of the average squared distance of points to the mean center:

$$s_d = \sqrt{\frac{\sum_{i=1}^{n} d_{ic}^2}{n}} \qquad (2.3)$$

where n is the number of points in the data, and d_{ic} is the distance from point i to the mean center.

With the aspatial version (i.e., the standard deviation), loosely speaking, the square root "undoes" the squaring, and thus, the standard deviation may be roughly interpreted as a quantity that is on the same approximate scale as the average absolute deviation of observations from the mean. Deviations from the mean are first squared (the deviations may be either positive or negative; squaring them yields a positive quantity). Taking the square root of the sum of these squared deviations then returns a quantity that is roughly on the same scale as the original deviations. However, in the spatial version, distances are always positive, and so a more interpretable and natural definition of standard distance would be to simply use the average distance of observations from the mean center (and in practice, the result would usually be fairly similar to that found using the preceding equation). Although Bachi's measure of standard distance is conceptually appealing as a spatial version of the standard deviation, it is not really necessary to maintain the strict analogy with the standard deviation by taking the square root of the average squared distance.

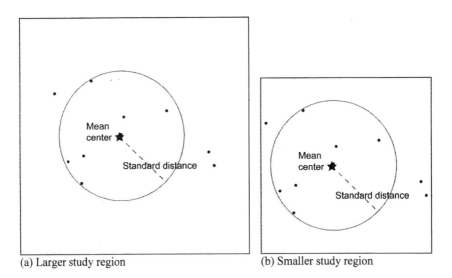

(a) Larger study region   (b) Smaller study region

FIGURE 2.2
Relative and absolute dispersion: (a) larger study region; (b) smaller study region.

2.5   Relative Standard Distance

One drawback of the standard distance measure described earlier is that it is a measure of *absolute dispersion*; it retains the units in which distance is measured. Furthermore, it is affected by the size of the study area. The two panels of Figure 2.2 show situations where the standard distance is identical, but clearly, the amount of dispersion about the central location, relative to the study area, is lower in panel (a).

A measure of relative dispersion may be derived by dividing the standard distance by the radius of a circle with area equal to the size of the study area (McGrew and Monroe 1993). This division makes the measure of dispersion unitless, and it standardizes for the size of the study area, thereby facilitating comparison of dispersion in study areas of different sizes. More formally, the relative standard distance is defined as

$$s_{d,rel} = \frac{s_d}{\sqrt{A/\pi}} \qquad (2.4)$$

where A is the size of the study area.

In the special case of a circular study area, the relative distance is $s_{d,rel} = s_d/R$, where R is the radius of the study area. For a square region, $s_{d,rel} = s_d\sqrt{\pi}/S$, where S is the length of the side of the square. Note that the maximum relative distance for a circle is 1; for a square, the maximum relative distance is $\sqrt{\pi}/2 = 1.253$.

2.6   Inferential Statistical Tests of Central Tendency and Dispersion

Simple inferential tests of central tendency and dispersion may be constructed using the preceding definitions, together with knowledge of the expected values and variability of the quantities under particular null hypotheses. Here, we give a few examples. These examples are elementary ones and are designed to give the reader a sense of how one can make the jump from descriptive measures to inferential tests. Their use is limited in practice due to the restrictive assumption that the background population is spatially homogeneous.

To take a simple case, suppose the study area is square, and the coordinates are scaled so that each side has length equal to one. Suppose we wish to test the null hypothesis that the true x-coordinate of the mean center is equal to 0.5 (of course, an identical test could also be carried out for the y-coordinate). A t-test may be employed; the observed value of the test statistic is $t_{obs} = (\bar{x} - 0.5)/(s/\sqrt{n})$, where $\bar{x}$ and s are the mean and standard deviations of the set of x-coordinates, respectively, and n is the number of sample locations. This test has $n - 1$ degrees of freedom, and assumes implicitly that the x-coordinates have a normal distribution. If the size of the sample is not too small, then this assumption is not critical. The value of t_{obs} is compared with a critical value observed from a t-table with $n - 1$ degrees of freedom and a prespecified probability of a Type I error (e.g., $\alpha = 0.05$). In addition, the probability of observing a test statistic that is more extreme than t_{obs} may also be found from the t-table (this is the p-value).

Tests of spatial dispersion may be carried out using the following facts (Eilon et al., 1971):

For a circle, the expected distance from the center to a randomly chosen point is $E[d] = 2R/3$. The variance of distances from the center to randomly chosen points is $V[d] = R^2/18$. For a square with sides of length S, the expected distance from the center to a randomly chosen point is

$$E[d] = (S/6)[\sqrt{2} + \ln(1 + \sqrt{2})] \approx 0.383S \tag{2.5}$$

and the variance of distances from the center to randomly chosen points is

$$V[d] = 2S^2/12 - (.383S)^2 \approx .02S^2 \tag{2.6}$$

More generally, for a rectangle with sides a and b, the expression is more complicated:

$$E[d] = \frac{r}{3} + \frac{a}{24}\left(\frac{a}{b}\ln(h_1) + \frac{b^2}{a^2}\ln(h_2)\right); \tag{2.7}$$

$$h_1 = \frac{r + b/2}{r - b/2}; \qquad h_2 = \frac{r + a/2}{r - a/2}; \qquad r = \frac{\sqrt{a^2 + b^2}}{2}$$

The associated variance is given by

$$V[d] = \frac{a^3b + ab^3}{12ab} - E[d]^2 \tag{2.8}$$

For example, with $a = 5$, and $b = 6$, the expected value of the distance from the center of the rectangle to a randomly chosen point is 2.11, and its variance is 0.63. A simple z-test associated with the null hypothesis that points are distributed randomly around the center may then be considered as follows:

$$z = \frac{\bar{d} - E[d]}{\sqrt{V[d]/n}} \tag{2.9}$$

where n is the number of points, and the quantity z is treated as a standard normal variate. Thus, although the standard distance is usually thought of as simply a descriptive measure of spatial data, it can also be used in an inferential context.

It is also possible to formulate tests based on the distribution of the square of the standard distance. It may be shown that, for a circle, the expected squared distance from the center to a randomly chosen point is $R^2/2$; the standard deviation of squared distances from the center to a randomly chosen point is $R^2/\sqrt{12}$. Similarly, for a square, the expected squared distance from the center to a randomly chosen point is $S^2/6$. The standard deviation of squared distances from the center to a randomly chosen point is $S^2/\sqrt{90}$. If we can assume, approximately, that the square of the standard distance has a normal distribution when points are randomly dispersed about the center (note that this assumption may not be tenable in practice), hypothesis tests can be based on z-scores. Thus, for a circle, a test of the null hypothesis of random dispersion about the center would entail the use of

$$z = \frac{s_d^2 - R^2/2}{R^2/\sqrt{12n}} = \frac{\sqrt{12n}\left(s_d^2 - R^2/2\right)}{R^2} = \sqrt{12n}\left(s_{d,rel}^2 - 0.5\right)$$

(2.10)

and for a square,

$$z = \frac{s_d^2 - S^2/6}{S^2/\sqrt{90n}} = \frac{\sqrt{90n}\left(s_d^2 - S^2/6\right)}{S^2} = \sqrt{90n}\left(\frac{s_d^2}{S^2} - \frac{1}{6}\right) = \sqrt{90n}\left(s_{d,rel}^2/\pi - 0.167\right)$$

(2.11)

where s_d and $s_{d,rel}$ are defined as Equations 2.3 and 2.4, respectively.

2.7 Illustration

The descriptive and inferential spatial statistics outlined earlier are now illustrated using the data in Table 2.1. This is a simple set of 10 locations as shown in Figure 2.3a; a small data set has been chosen deliberately

TABLE 2.1

Coordinate Locations for 10 Sample Points

x	y
0.8616	0.1781
0.1277	0.4499
0.3093	0.5080
0.4623	0.3419
0.4657	0.3346
0.2603	0.0378
0.6680	0.3698
0.2705	0.1659
0.1981	0.1372
0.8891	0.1192

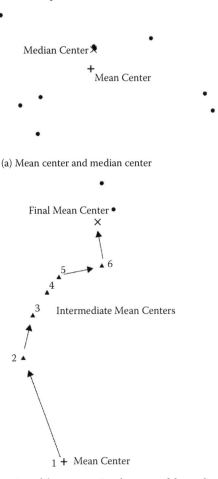

(a) Mean center and median center

(b) Illustration of the computational process of the median center

FIGURE 2.3
Mean center, median center, and computational process of the median center: (a) mean center and median center; (b) illustration of the computational process of the median center.

to facilitate the derivation of the quantities by hand. The study area is assumed to be a square with the length of each side equal to one. In this example, we assume implicitly that there are equal weights at each location (or equivalently, that one individual is at each location).

The mean center is (0.4513, 0.2642), and this is found by simply taking the mean of each column ("+" symbol in Figure 2.3a). The median center is (0.4611, 0.3312) ("×" symbol in Figure 2.3a). Accuracy to three digits is achieved after 33 iterations. The first few iterations of Equations 2.2 are shown in Table 2.2 and Figure 2.3b. It is interesting to note that the

TABLE 2.2

First Few Iterations in Calculation
of Median Center

x	y
0.4512	0.2642
0.4397	0.2934
0.4424	0.3053
0.4465	0.3116
0.4499	0.3159
0.4623	0.3191
.	.
.	.
.	.
.	.
0.4611	0.3312

approach to the y-coordinate of the median center is monotonic, whereas the approach to the x-coordinate is a damped harmonic.

The sum of squared distances to the mean center is 0.8870; note that this is lower than the sum of squared distances to the median center (0.9328). Similarly, the sum of distances to the median center is 2.655, and this is lower than the sum of distances to the mean center (2.712).

The standard distance is 0.2978 (which is the square root of 0.8870/10); note that this is similar to the average distance of a point from the mean center (2.712/10 = 0.2712).

Testing the hypothesis that the points are randomly distributed about the center of the square results in a z-score of $z = (0.2712 - 0.383)/(\sqrt{.02/10}) = -2.500$ (see Equations 2.5, 2.6, and 2.9). Because this is less than the critical value of 1.96 associated with a two-sided test using $\alpha = 0.05$, we reject the null hypothesis and conclude that points are clustered to a greater degree about the center than would be expected under the null hypothesis of random dispersion about the center. A similar conclusion is obtained using Equation 2.11 because

$$z = \sqrt{90(10)}\, \frac{0.0887 - 0.1667}{1} = -2.35 \tag{2.12}$$

2.8   Angular Data

Some applications make use of angular data. For example, not only may distance to a hazardous waste site be important in determining health risk but the direction of the prevailing wind may be important as well. Angular data pose special issues when one attempts description and inference. To see this, take the simple case in which we are interested in wind direction and use $0°$ as our reference point for the northerly direction. Suppose now that we collect two measurements: $1°$ and $359°$. Clearly, the average direction is north (each observation differs by only $1°$ from north), but if we take a simple arithmetic average of the two observations, we find that the mean direction is due south ($[1 + 359]/2 = 180°$)! It is evident that alternative approaches are necessary.

The following steps may be used to find out the mean and variance of a set of angular observations:

1. Find the mean of the cosines (C) and sines (S) of the observations.
2. Find $R = \sqrt{C^2 + S^2}$.
3. The mean angle A is the angle that has $\cos A = C/R$, and $\sin A = S/R$. That is,

$$A = \sin^{-1}(S/R); \quad A = \cos^{-1}(C/R).$$

The variance of the angles, termed the *circular variance*, is equal to $1 - R$. When R is high, $1 - R$ (the circular variance) is low, and there is not much variability in angles. Again, the circular variance refers to the dispersion of *angles*.

The Rayleigh test is a test of the null hypothesis that angles are distributed randomly around the circle (Mardia 1972, p. 133). The test compares R with a critical value (e.g., given in Appendix 2.5 of Mardia). If R is greater than the critical value, this implies that the null hypothesis should be rejected in favor of the alternative that the angles are clustered.

Illustration 1: Let us consider a very simple data set that consists of the following five angles (Figure 2.4a).

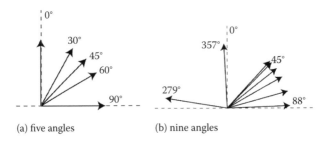

(a) five angles   (b) nine angles

FIGURE 2.4
Examples of angular data: (a) five angles; (b) nine angles.

Angle (Degrees)	Cos	Sin
0	1.0000	0.0000
30	0.8660	0.5000
45	0.7071	0.7071
60	0.5000	0.8660
90	0.0000	1.0000

$n = 5$; C = mean of cosines = 0.6146; S = mean of sines = 0.6146; R = sqrt $(C^2 + S^2) = 0.8692$; mean angle = A; $\cos A = C/R = 0.7071$; $\sin A = S/R = 0.7071$; $A = 45.0°$; circular variance = $1 - R = 0.1308$.

The critical value of R with $n = 5$ and $\alpha = 0.05$ is 0.754. Because 0.8692 >0.754, reject the null hypothesis, and conclude that there is less variability than one would expect in the angles had they been randomly distributed.

Illustration 2: (Figure 2.4b; from Mardia [1972])

Angle (Degrees)	Cos	Sin
43	0.7314	0.6820
45	0.7071	0.7071
52	0.6157	0.7880
61	0.4848	0.8746
75	0.2588	0.9659
88	0.0349	0.9994
88	0.0349	0.9994
279	0.1564	−0.9877
357	0.9986	−0.0523

$n = 9$; C = mean of cosines = 0.4470; S = mean of sines = 0.5529; R = sqrt $(C^2 + S^2) = 0.7110$; mean angle = A; $\cos A = C/R = 0.6287$; $\sin A = S/R = 0.7776$; $A = 51.0°$; circular variance = $1 - R = 0.2890$.

The critical value of R with $n = 9$ and $\alpha = 0.05$ is 0.569. Because 0.7110 >0.569, reject the null hypothesis and conclude that there is less variability than one would expect in the angles had they been randomly distributed.

2.9 Characteristics of Spatial Processes: First-Order and Second-Order Variation

First-order variation refers to the variations in the mean of a process over space; for example, the mean incidence of lung cancer over space may vary due to spatial variation in radon concentrations. *Second-order* variation in a spatial process results from spatial dependency. For example, the spatial distribution of influenza is at least partly attributable to contagion and diffusion processes. Areas with high rates tend to be located near other areas with high rates because of the second-order characteristics of the influenza process. Thus, first-order effects result from spatial variation in exogenous explanatory variables (for example, the mean of house prices in a study area varies with accessibility to transit lines), whereas second-order effects result from communication, transmission, and/or interaction processes (Bailey and Gatrell 1995, p. 32). Disease distributions can be a function of first-order effects (e.g., location of hazardous waste) or second-order effects (e.g., communicable disease), or some combination of the two.

Stationary processes are those in which the mean and variance do not vary with location; in addition, the covariance of values observed at distinct locations is dependent on distance and direction only. *Isotropic processes* have a more restrictive definition—they are stationary processes in which the covariance is only dependent on distance (and not direction).

Bailey and Gatrell (1995, p. 35) make the important point that "concepts such as stationarity and first- or second-order effects are artifacts of the modeler and not reality. In practice, the effects are confounded in observed data, and the distinction between them is difficult and, ultimately, to some extent arbitrary. Both types of effect can give rise to similar spatial patterns." Thus, when a cluster of high-risk locations is found on a map, it is difficult, if not impossible, to tell whether it has arisen due to either spatial variation in the mean (e.g., environmental effects), or to second-order effects (e.g., a contagious process).

How can we distinguish spatial dependence in a homogeneous environment from spatial independence in a heterogeneous environment? One idea is to subtract out the first-order effects and look for second-order effects (this is, for example, the general strategy used in Cressie's *Statistics for Spatial Data* 1993). An alternative that we will return to is to do the opposite—subtract out second-order effects and look for first-order effects. However, in either case the two effects are comingled, and this is not as straightforward as it sounds. In subtracting out first-order effects, we may actually be removing some of the second-order effects. In subtracting out second-order effects, we may be including some first-order effects in the subtraction.

2.10   Kernel Density Estimation

The purpose of kernel density estimation is to estimate the intensity of a point process at given locations. The intensity of a point process at a given location is defined as the *limiting density of points* (i.e., number of points per unit area), as the area around the given location becomes smaller and smaller. Typically, kernel density estimation is carried out on a set of grid points covering the study area. Based on the intensity estimates made at these points, an intensity surface is generated.

The first step of the process is to place a kernel at a grid point i; each observed point location, j ($j = 1, ..., n$), is then assigned a weight according to the kernel function $k(d_{ij})$, which is a function of the distance from grid point i to point location j. The intensity estimate at i is the sum of the individual contributions made from each observed point j:

$$\hat{\lambda}_i = \sum_{j=1}^{n} k(d_{ij}) \tag{2.13}$$

Figure 2.5 provides a simple one-dimensional example to illustrate that an observed point j_1 that is closer to the grid point i is assigned a higher weight than the farther point j_2 whose assigned weight is in this case equal to zero.

Many choices are available for the kernel function; a common one is the quartic kernel:

$$k(d_{ij}) = \frac{3}{\pi\tau^2} \left(1 - \frac{d_{ij}^2}{\tau^2}\right)^2 ; \quad d_{ij} < \tau \tag{2.14}$$

where τ is the "bandwidth" of the kernel. Larger values of τ imply more smoothing; all points within a distance τ of the grid point receive a non-zero weight in the intensity estimator.

The estimator $\hat{\lambda}_i$ is, preferably, adjusted for edge effects by dividing it by the proportion of the kernel's volume that lies within the study area R:

$$\hat{\lambda}_{i,edge} = \frac{\hat{\lambda}_i}{\int_{all\ l \in R} k(d_{il})} \tag{2.15}$$

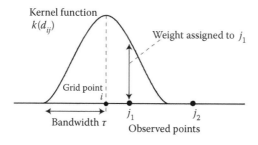

FIGURE 2.5
Basic concepts of kernel density estimation.

A small bandwidth uses only local information in the vicinity of a grid point; the resulting estimate of intensity will have little bias but will have a high degree of uncertainty because it will be based on little information. On the other hand, a large bandwidth results in an estimate of intensity that has less uncertainty but one that will have a greater degree of bias because more nonlocal information is being used.

Consequently, various suggestions have been made for an "optimal" bandwidth that represents a compromise in this trade-off between bias and uncertainty. Bailey and Gatrell (1995) note the suggestion of $\tau = 0.68\ n^{-0.2}$ made for unit square study areas. Waller and Gotway (2004) note that, for a Gaussian kernel, Scott (1992) recommends a bandwidth defined by the standard deviation of the Gaussian function equal to $\hat{\sigma}n^{-1/6}$, where $\hat{\sigma}$ is the standard deviation of the x- or y-coordinates. "Adaptive" kernels may also be used, in which the bandwidth used in a given study varies from location to location depending on the density of points. Thus, larger bandwidths are used in regions of low density, and smaller bandwidths are used in regions with high point densities.

Example

Using the Humberside leukemia case data and a bandwidth of 100, a density surface is estimated as shown in Figure 2.6. A smaller bandwidth (20) leads to the more detailed image shown in Figure 2.7.

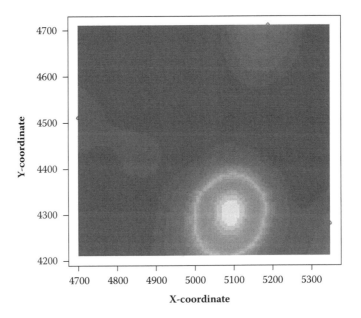

FIGURE 2.6

Kernel density estimates for Humberside leukemia data: bandwidth = 100.

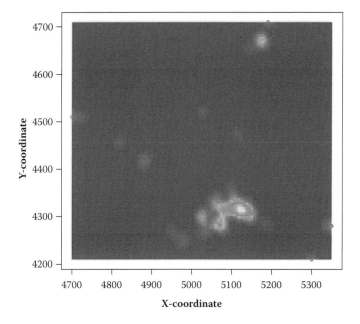

FIGURE 2.7
Kernel density estimates for Humberside leukemia data: bandwidth = 20.

2.11 K-Functions

The *K*-function is a measure of the second-order characteristics of a point process. It is defined as the number of events within a distance *h* of an arbitrary event, divided by the mean number of events per unit area. The expected value of *K*(*h*) in a random pattern is equal to $\lambda \pi h^2 / \lambda = \pi h^2$, where λ is the mean number of events per unit area.

An estimate of *K*(*h*) is based on a count of the number of pairs of points that are found within a distance *h* of one another. More formally,

$$\hat{K}(h) = \frac{1}{\hat{\lambda}^2 R} \sum_{i}^{n} \sum_{j \neq i}^{n} I_{ij}(h) = \frac{R}{n^2} \sum_{i}^{n} \sum_{j \neq i}^{n} I_{ij}(h) \qquad (2.16)$$

where *R* is the area of the study region, *n* is the number of points observed in *R*, and $I_{ij}(h)$ is equal to one if points *i* and *j* are separated by a distance less than *h*, and is equal to zero otherwise. In addition, the estimate of average intensity has been replaced by $\hat{\lambda} = n/R$.

It is important to account for "edge effects" because we would expect the count of points within a distance *h* of a point to be relatively small if the point was near the border of the study area. To account for such effects,

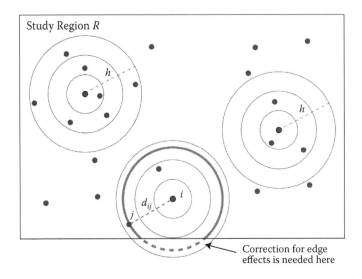

FIGURE 2.8
Basic concepts of the *K*-function and associated edge-correction.

the $I_{ij}(h)$ term is divided by w_{ij}, defined as the proportion of the circumference of a circle centered on *i* and passing through *j* that lies within the study region (i.e., the bold gray part of the circle indicated in Figure 2.8).

The term w_{ij} is the conditional probability that a point falls within the study region, given that it is a distance d_{ij} from point *i*. The edge-corrected estimate is thus

$$\hat{K}_{edge}(h) = \frac{R}{n^2} \sum_{i} \sum_{j \neq i} \frac{I_{ij}(h)}{w_{ij}} \tag{2.17}$$

A range of *h* values is generally examined by means of a plot of $\hat{K}(h)$ versus *h*. For a random pattern, the expected value of $K(h)$ is πh^2 so that the plot will be an upward-sloping quadratic curve as shown in Figure 2.9. If points are more clustered than random, $\hat{K}(h)$ will lie above this curve because there will be a higher number of pairs of points within a distance *h* of one another (i.e., $\hat{K}(h) > \pi h^2$). Similarly, a value of $\hat{K}(h) < \pi h^2$ implies that points are more spread out than random.

To assess statistical significance, one can simulate spatial randomness by locating *n* points at random in the study area. The *K*-function is found for the simulated pattern, and this is repeated a large number of times. All of the resulting $\hat{K}(h)$ values are ordered and, for each *h* value among a set of *h* values, the 5th and 95th percentiles of the simulated $\hat{K}(h)$ values are plotted. This results in a set of simulation envelopes that surround the expected $K(h) = \pi h^2$ curve, and they correspond to a significance level of 10%.

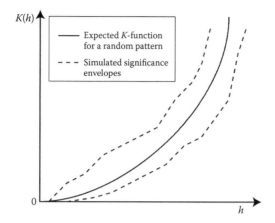

FIGURE 2.9
Expected value of the K-function for a random pattern and significance envelopes.

If the observed K-function falls outside these envelopes, it implies that a test of the null hypothesis of randomness at a scale *h* would be rejected. It is important to note that the envelopes are useful in testing the null hypothesis at a *particular* spatial scale. If a range of scales is examined, as is often the case, it would be even more likely than the nominal value of $\alpha = 0.10$ in this instance (corresponding with the use of the 5th and 95th percentiles) that a plot of $\hat{K}(h)$ versus *h* would extend beyond the envelopes for *at least some* value of *h*.

Example

For the Humberside leukemia data, we see from Figure 2.10 that there is significant spatial clustering of cases because the observed K-function lies well outside the simulation envelopes. This clustering occurs at all spatial scales.

2.12   Differences and Ratios of Kernel Density Estimators

Let $\hat{\lambda}_i^{case}$ be the intensity of cases at point *i*, and $\hat{\lambda}_i^{control}$ be the intensity of controls at point *i*. Kelsall and Diggle (1995) suggest that a comparison be based on the natural logarithm of the ratio of these two:

$$r_i = \ln\left(\hat{\lambda}_i^{case} / \hat{\lambda}_i^{control}\right)$$

(2.18)

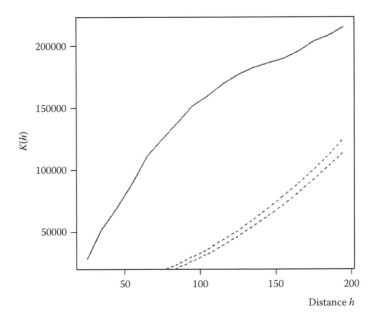

FIGURE 2.10
K-function for Humberside leukemia data.

Significance is often assessed by randomly relabeling cases and controls, computing *r*, repeating this many times, and then noting where the observed value of r_i falls in relation to the tails of the distribution of *r* constructed from the many relabeled simulations.

Example

The ratio of case to control densities for the Humberside leukemia data, using a bandwidth of 100, is shown in Figure 2.11.

Because the denominator of the ratio of kernel densities can possibly be quite small, making the ratio itself large and potentially unstable, an alternative is to use the standardized difference between case and control densities. This allows one, for example, to identify areas with differences between case and control densities exceeding two standard deviations (Bowman and Azzalini 1997). Han et al. (2005) indicate that "[T]he standardized difference between case and control densities is obtained by taking the square root of the case density minus the square root of the control density, and dividing by the standard deviation of the difference between densities." Thus,

$$\frac{\sqrt{\lambda_i^{case}} - \sqrt{\lambda_i^{control}}}{\sqrt{var\left(\sqrt{\lambda_i^{case}} - \sqrt{\lambda_i^{control}}\right)}} \tag{2.19}$$

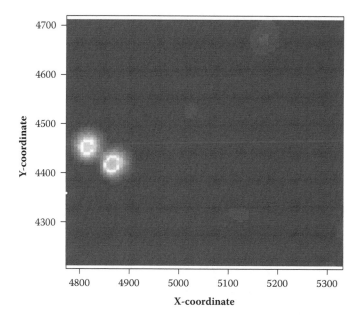

FIGURE 2.11
Ratio of case to control densities for Humberside leukemia data with bandwidth = 100.

where

$$\text{var}\left(\sqrt{\lambda_i^{case}} - \sqrt{\lambda_i^{control}}\right) \approx \frac{1}{4h}\left(\frac{1}{n_1} + \frac{1}{n_2}\right)\int k_h^2(x)dx \qquad (2.20)$$

and n_1 and n_2 are the number of cases and controls, respectively, and k is the kernel density function. Both the kernel density function and the kernel density estimators are scaled to integrate to one. An example of this approach is given by Han et al. (2005) who studied the geographic pattern of breast cancer in western New York State. They use contours to map areas where the standardized case density exceeds the standardized control density (and the standardized difference is greater than two).

Because many geographic locations are examined simultaneously, it is not correct to use 2 (or more precisely, 1.96, assuming a Type I error probability of .05) because some locations should be expected to exceed this value by chance alone. A more appropriate critical value is one that will yield a 0.05 chance that the maximum difference in case and control is exceeded. By randomly assigning case and control status to the fixed locations of cases and controls, this critical value can be established; in the present example, it was found to be 3.56. Areas with standardized density differences exceeding this value are also shown in the figure (Figure 2.11).

2.13　Differences in *K*-Functions

Comparisons of spatial patterns are often desirable. For example, it may not be of particular interest to simply characterize the spatial pattern of disease cases because we might expect it to be clustered (e.g., due to clustering of the population). Rather, we would like to compare the nature of the clustering of cases with either the clustering of controls or the clustering of the population.

Let $K_1(h)$ and $K_2(h)$ represent the *K*-functions for cases and controls, respectively. The quantity $K_1(h) - K_2(h)$ may be used to compare the two patterns. Note that the expected value of the difference of the two *K*-functions is zero under the null hypothesis that the cases and controls have spatial clustering tendencies of equal strength.

A positive value indicates spatial clustering of cases that is greater than the spatial clustering of controls; similarly, a negative value indicates that the controls are relatively more clustered than the cases.

One approach to assessing statistical significance is to randomly relabel the cases and controls, derive the *K*-functions for each, and compute the difference in *K*-functions. If this is repeated many times, the 5th and 95th percentiles found for different values of *h* may be used to establish

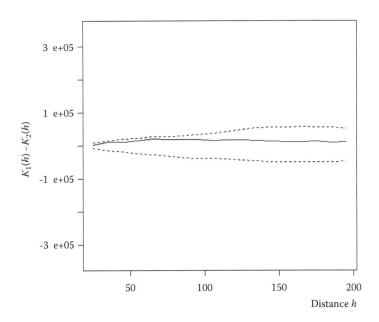

FIGURE 2.12
Difference between *K*-functions for cases and controls for Humberside leukemia data.

simulation envelopes. If the observed difference in *K*-functions lies outside this envelope, it implies that the null hypothesis of no difference in spatial patterns should be rejected.

Example

Figure 2.12 shows the difference between case and control *K*-functions for the Humberside leukemia data. Note that the observed line remains inside the envelopes, indicating that there is no significant difference in the degree to which the case and control patterns cluster.

3

Global Statistics

3.1 Introduction

Global spatial statistics summarize, in a single quantity, the degree to which an observed spatial pattern deviates from a specified null hypothesis. Examples of null hypotheses in this context include those of complete spatial randomness, and those in which observed regional frequencies are consistent with a set of prespecified regional expectations for those frequencies. Observed values of the global statistics are compared with the distribution of statistics that is expected under the null hypothesis, and this comparison leads to acceptance or rejection of the null hypothesis, along with a p-value indicating the likelihood of the observed statistic under the null hypothesis. Global statistics are limited in the sense that they do not give a direct indication of the size and location of regions that are inconsistent with the null hypothesis.

In this chapter we review several global statistics that have been developed to test the null hypothesis of spatial uniformity in risk. We begin with two of the earliest methods: the quadrat method and the nearest neighbor statistic. Both of them were developed in the field of ecology and were designed to test the null hypothesis that the geographic distribution of species did not differ from complete spatial randomness. These tests, as originally designed, are ultimately quite limiting when it comes to many applications such as disease clustering; because populations are not spatially random, there is no reason to suspect that the distribution of disease should be spatially random. Instead, statistical tests are required that account for the geographic distribution of population, and more generally, other covariates as well. For example, disease incidence may vary with age, income, occupation, and a host of other factors. We often wish to find any geographic clusters that exhibit excess risk—*after* all of these known factors have been accounted for.

The quadrat test may be so modified to account for covariates; it is based on the observed and expected numbers of cases that occur in each of a number of subregions that comprise a study area. We emphasize this test in this chapter because it forms the foundations for many of the other tests that follow.

We also look at other tests based on areal data—Moran's I, Oden's I_{pop}, and Geary's C. Of these, Moran's I is by far the most widely used. We then introduce Tango's global statistic and show how the quadrat approach may be combined with an approach similar to Moran's *I* into a single, Tango-like, spatial chi-square test. Finally, we review the Cuzick–Edwards test as an example of a global test based on data containing the precise locations of points; this test does not have some of the limitations of the original nearest neighbor statistic.

3.2   Nearest Neighbor Statistic

The nearest neighbor statistic was developed primarily in the context of applications to the spatial distribution of plant populations. In addition to the early work of Skellam (1952), the work of Clark and Evans (1954) was instrumental in popularizing the approach. The latter authors derived the expected mean and variance of the distribution of distances between points and their nearest neighboring points in a random pattern. This allows for a test of spatial randomness through the use of the *z*-score:

$$z = \frac{\bar{r}_0 - r_e}{s_e} \tag{3.1}$$

where $\bar{r}_0$ is the mean observed distance between each point and its nearest neighboring point. Furthermore, r_e and s_e are the expected value and standard deviation associated with distances between points and their nearest neighboring points in a spatially random set of points. These quantities are determined through

$$\bar{r}_0 = \frac{\sum_{i=1}^{n} d_i}{n}; \quad r_e = \frac{1}{2\sqrt{n/A}}; \quad s_e = \frac{.261}{\sqrt{n^2/A}} \tag{3.2}$$

where d_i is the distance from point *i* to its nearest neighbor, *n* is the number of points in the study area, and *A* is the size of the study area. Nearest neighbors may in some cases be reflexive (i.e., they are nearest neighbors of each other), but they are not necessarily so.

The nearest neighbor statistic itself is often expressed as a ratio of the observed mean distance to the expected distance $(\bar{r}_0/r_e)$. Ratios less than one indicate a tendency toward clustering, and ratios greater than one indicate a tendency toward uniformity or regularity.

The nearest neighbor statistic is subject to boundary effects. In some cases, the nearest neighbor of a point in the study area happens to lie outside of the study area. If the nearest neighbor is restricted to be a point inside of the study area, the observed mean distance between nearest

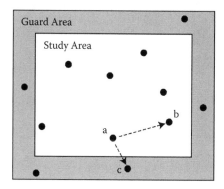

FIGURE 3.1
Guard area method to avoid boundary effects.

neighbors will tend to be higher than the value of r_e in a randomly distributed set of points. For example, in Figure 3.1, when only points in the study area are considered in the analysis, the nearest neighbor distance for point a will be the distance from a to b, which is much longer than its true value, that is, the distance from a to c. A solution is to establish a "guard" area around the study area; then, for each point in the study area, one finds the distance to the nearest neighboring point, regardless of whether the neighboring point lies in the study area or in the guard area. The nearest neighbor distance for point a in Figure 3.1 will actually be measured as the distance from a to c. Note that points in the guard area are used only to measure distances to them, not from them.

Because nearest neighbor distances are employed (and distances to other points are ignored), the usefulness of the approach in detecting departures from spatial randomness is confined to situations in which clustering occurs on relatively small spatial scales.

3.2.1 Illustration

The nearest neighbor distances for the 10 points located in a unit square study area (see Table 2.1 and Figure 2.3a) are given in Table 3.1. The mean distance to nearest neighbors is equal to 0.1006. The expected distance between neighbors in a random pattern is $1/(2\sqrt{10/1}) = 0.1581$. The nearest neighbor ratio is $0.1006/0.1581 = 0.631$, indicating a tendency for the points to cluster because the ratio is less than one. A test of the null hypothesis of spatial randomness confirms that the clustering is significant at the 5% significance level, and is unlikely to have occurred by chance alone:

$$z = \frac{0.1006 - 0.1581}{0.26/\sqrt{10(10)}} = -2.21 \tag{3.3}$$

TABLE 3.1

Nearest Neighbor Distances for 10 Sample Points

x	y	Distance to Nearest Neighbor
0.8616	0.1781	0.065
0.1277	0.4499	0.191
0.3093	0.5080	0.191
0.4623	0.3419	0.008
0.4657	0.3346	0.008
0.2603	0.0378	0.117
0.6680	0.3698	0.205
0.2705	0.1659	0.078
0.1981	0.1372	0.078
0.8891	0.1192	0.065

This corresponds to a *p*-value of .014. To explore boundary effects, a new set of 10 points were randomly located in the study area, and the mean distance to nearest neighbors was noted (there was no guard area applied here). One thousand repetitions of this resulted in a mean distance between nearest neighbors of 0.1845—notably larger than the expected value of 0.1581 determined by the equation for expected nearest neighbor distance. Using this in the test of the null hypothesis would have led to a result in the opposite direction, implying a tendency toward regularity though not statistically significant:

$$z = \frac{0.1845 - 0.1581}{0.26/\sqrt{10(10)}} = 1.02 \tag{3.4}$$

3.3   Quadrat Methods

Quadrat methods for detecting global departures from randomness are based on a set of cells, or quadrats, that are placed over the study area. They are based on a comparison of the observed and expected counts in each cell. Quadrat methods may be classified into two principal alternative approaches that differ with respect to whether the total number of points falling into the study area is either known (conditional approach) or unknown (unconditional approach). We now illustrate these with the following example.

Suppose that a 4×4 grid is overlaid on a study area, and the resultant set of counts (e.g., of disease cases) in each grid cell is given as

2 4 3 4
0 2 1 4
2 2 1 2
4 0 0 1

3.3.1 Unconditional Approach

Adopting the approach of Student (1907), we can determine the goodness of fit of these data to a Poisson distribution. The mean count observed in a cell is equal to two (i.e., the total number of points, 32, divided by the number of cells, 16); this is an estimate of the intensity of the process at the scale of the cell. Assuming that the counts in all cells follow Poisson distributions with mean two, we first calculate the expected proportion of cells with counts of 0, 1, 2, ... , etc. These are determined by using the Poisson distribution:

$$\Pr(X = x) = \frac{e^{-\lambda}\lambda^x}{x!} \tag{3.5}$$

The result is the second column of Table 3.2. By multiplying each entry in column 2 by the total number of cells (16), we arrive at the expected number of cells with a particular count. This may then be compared with the observed number of cells with that count. Assessment of statistical significance is carried out by using a χ^2 goodness-of-fit statistic:

$$\chi^2_{obs} = \sum_{i=1}^{k} \frac{(O_i - E_i)^2}{E_i} \tag{3.6}$$

TABLE 3.2

Observed and Expected Counts: Unconditional Approach to Quadrat Analysis

Count	Expected Proportion of Cells	Expected Number of Cells with Count	Observed Number of Cells with Count
0	0.135	2.17	3
1	0.271	4.33	3
2	0.271	4.33	5
3	0.180	2.89	1
4	0.090	1.44	4
5+	0.053	0.84	0

where O_i and E_i are, respectively, the observed and expected counts of cells in category i, and k is the number of categories. A common rule of thumb is that the expected count in each category should be equal to five or more, but Koehler and Larntz (1980) have found that this is too stringent (this is discussed further in subsection 3.3.3). In the preceding example, if we combine the last two categories (so that the expected count is 2.28, and the observed count is 4), we find that $\chi^2_{obs} = 3.36$. This statistic is to be compared with a critical value of χ^2 with degrees of freedom equal to $k-1$. In our example, if we use $\alpha = 0.05$, this implies that we use a critical value of $\chi^2_{.05,4df} = 11.07$. Because $\chi^2_{obs} < \chi^2_{.05,4df}$, we accept the null hypothesis and conclude that the data do not deviate significantly from what one would expect if random counts (having a mean of 2) were assigned to each cell. Furthermore, use of the chi-square table reveals that $p = .50$, implying that when the null hypothesis is true, 50% of the time we would expect an even more extreme value of χ^2 than the one observed. Monte Carlo simulation was also carried out; 1000 simulations of the null hypothesis achieved by filling each cell randomly using a Poisson distribution with a mean of two led to a similar p-value of .511.

3.3.2   Conditional Approach

The conditional approach adopts the perspective that, in repeated sampling, the total number of points is always equal to n. Simulations of the null hypothesis are carried out *conditional* on the total being equal to n; the n points are assumed to be randomly distributed across the m cells. This is in contrast with the unconditional approach, where the total number of points varies from one realization of the null process to the next. A more commonly used alternative method to test for departures from randomness is to use the variance-mean ratio (VMR). If the null hypothesis of randomness is true, the quantity $(m-1)$ VMR will have a χ^2 distribution with degrees of freedom equal to $m-1$, where m is the number of subregions. We begin by calculating

$$VMR = \frac{\sum_{i=1}^{m}(O_i - \bar{O})^2}{(m-1)\bar{O}} \tag{3.7}$$

where $\bar{O}$ is the mean count per cell, that is, n/m. Then,

$$\chi^2_{obs} = (m-1)VMR = \frac{\sum_{i=1}^{m}(O_i - \bar{O})^2}{\bar{O}} \tag{3.8}$$

In the present example, the variance (i.e., $\sum_{i=1}^{m}(O_i - \bar{O})^2/(m-1)$) is equal to 32/15, and the mean $\bar{O}$ is equal to two; this leads to VMR = (32/15)/2

= 16/15 = 1.067. χ^2_{obs} is equal to 16, and this is to be compared with a critical value of $\chi^2_{.05,15df}$ = 25.0. Because $\chi^2_{obs} < \chi^2_{.05,15df}$, we accept the hypothesis of randomness.

Furthermore, we find from a χ^2 table with 15 degrees of freedom that $\Pr(\chi^2 > 16) = 0.38$. The *p*-value of .38 implies that 38% of the time we could expect a statistic more extreme than the one we observed, assuming the null hypothesis to be true.

This again was checked using 1000 Monte Carlo repetitions of the null hypothesis—each time randomly allocating 32 points to the 16 cells (i.e., simulation of the null hypothesis is *conditional* on the fact that there are 32 points in the study area). Of the 1000 tests, 41.7% had a more extreme test statistic than the observed statistic. Note that this is close to the tabled value of 38%.

It is important to note that Equation 3.8 may be generalized to the situation in which the expected count is not necessarily the same in each cell:

$$\chi^2_{obs} = \sum_{i=1}^{m} \frac{(O_i - E_i)^2}{E_i} \tag{3.9}$$

where E_i is the expected count in cell *i*. This is simply the common chi-square goodness-of-fit test.

3.3.2.1 Example 1: Leukemia in Central New York State

The chi-square goodness-of-fit test, when applied to the $n = 592$ leukemia cases for central New York State, yields a value of $\chi^2 = 430$. As the number of regions is $m = 281$, the associated degrees of freedom is $m - 1 = 280$. Using a Type I error probability of $\alpha = 0.05$, the critical value of χ^2 with 280 degrees of freedom is 320. Because the observed value of 430 exceeds this critical value, we reject the null hypothesis that cases are distributed randomly among the population.

This same example may be illustrated by using *GeoSurveillance 1.1*, introduced in Chapter 1, subsection 1.7.1.1. To begin, choose Cluster Detection, Retrospective Test, Score and M statistic, and Point (Text file). In the Cluster Detection—Score and M test panel, click on the Load Data button. Then, in the Data Format panel, choose Obs-Exp, click on Continue, and load the file NYTract_ObsExp.txt. Then choose Adjusted-Poisson under Variables and change the bandwidth to zero. Click Run. The results are given in the box labeled Global Pattern Summary.

3.3.2.2 Example 2: Sudden Infant Death Syndrome in North Carolina

The expected number of sudden infant death syndrome (SIDS) deaths for the period 1974–1978 for a county in North Carolina is first derived by multiplying the total number of SIDS deaths ($n = 667$) by the fraction of statewide births occurring in that county. The observed chi-square statistic of 225.6 is to be compared with a critical value based on $m - 1 = 99$ degrees of freedom because there are $m = 100$ counties. As the observed value of 225.6 is greater than the critical value of $\chi^2_{\alpha=0.05,99df} = 123.2$, the null hypothesis is rejected.

When the degrees of freedom are large (greater than about 30), an alternative is to use the quantity $\sqrt{2\chi^2} - \sqrt{2(df)-1}$, which has a standard normal distribution. Thus, $\sqrt{2(225.6)} - \sqrt{2(99)-1} = 7.21$ is to be treated as a standard normal deviate, and clearly this is highly significant, indicating that it is extremely unlikely that the 667 SIDS cases could have been allocated to counties randomly, based on the number of births in each county.

In Chapter 4, Section 4.9, we consider the individual, local contributions of each county to this global chi-square value.

To carry out this example using *GeoSurveillance 1.1*, choose Cluster Detection, Retrospective Test, Score and M statistic, and Polygon (ESRI Shapefile). Load the file NCSIDS.shp. Then leave the Pop-Case (the default) option selected, and using the drop-down menus, choose BIR74 for Pop and SID74 for Case. Choose NAME for Primary ID and click OK. Under Bandwidth, choose Single value and set it to zero. Then click Run. Results are shown in the Global Pattern Summary box.

3.3.2.3　Example 3: Lung Cancer in Cambridgeshire

Haining (2003) gives the observed and expected number of lung cancer cases for 157 wards in Cambridgeshire, England. There are a total of 344 observed and 344 expected cases across the 157 wards. The χ^2 statistic is 154.75; the critical value of a χ^2 variable using $157 - 1 = 156$ degrees of freedom, and $\alpha = 0.05$ is 186, and thus, the null hypothesis that the observed counts are consistent with the expected counts (which in turn were based on population and age structure) is not rejected. The *p*-value associated with the test is equal to .51.

There are a number of open questions regarding these examples.

- What are the magnitudes of the contributions to the global statistic that are made by individual regions?
- For those cases where the global statistic is significant, can we identify specific regions that are prime contributors?
- When the global statistic is not significant, are there still individual regions that may have significant departures from expected values?

We will examine these and other questions in subsequent chapters.

3.3.3   Minimum Expected Frequencies

The usual guideline of an expected frequency of five cases per cell may often be violated, and this could call the conclusions into question. However, Koehler and Larntz (1980) find that this guideline is too stringent; they recommend that for symmetrical null hypotheses (where all expected values are equal), the chi-square statistic is adequate as long as $n^2/k \geq 10$ (where $k \geq 3$ and $n \geq 10$). For example, with the North Carolina SIDS data (subsection 3.3.2.2), there are many counties where the expected frequency is less than five, but Koehler and Larntz's conditions are easily satisfied ($n^2/k = 667^2/100 \gg 10$), though it should be noted that the null hypothesis is not symmetric here.

Koehler and Larntz also observe that the chi-square approximation is liberal when many expected values are less than one (although they base this conclusion on alternatives whose asymmetry is much more extreme than the present case).

To illustrate the potential effects of small expectations on the distribution of the χ^2 statistic under the null hypothesis, simulations were carried out. For both the North Carolina SIDS data and the New York leukemia data, n cases were distributed across m regions, using the multinomial probabilities that are found by dividing regional expectations by the total of expectations across all regions.

For the North Carolina SIDS data, $n = 667$ and $m = 100$; 125 was found to be the simulated 95th percentile of the χ^2 distribution under the null hypothesis that all counties had the same probability that a birth resulted in SIDS. Use of the chi-square table would have led to a choice of 123.2 for the critical value, using 99 degrees of freedom. Thus, the use of the table is slightly liberal—too many null hypotheses would be rejected.

For the leukemia data, $n = 592$ and $m = 281$; the 95th percentile of the simulated distribution of chi-square (based on simulations of the null hypothesis, where the 592 cases are assigned to the 281 census tracts according to their individual multinomial probabilities, which are in turn determined by population size) was 330. Use of the chi-square table would have led to a choice of 320 for the critical value; again, this is a value that is a bit too low, and would lead to over-rejection of the null hypothesis.

3.3.4   Issues Associated with Scale

The results of tests based on the aggregation of data into regions or quadrats will, of course, be dependent on the size of the spatial units. If clustering occurs on a smaller scale, it may remain undetected because the data have been aggregated into regions larger than the scale on which the clustering is operating. If data on location are sufficiently detailed, disaggregation to smaller spatial units could be undertaken. However, this is often not possible, and the regional aggregation imposes a lower limit on the scales of pattern that may be detected. For example, if cancer data are only available at the county scale, it will clearly not be possible to detect within-county clusters!

If clustering exists on a larger scale, it is important to account for the fact that deviations from expected values in one cell will be correlated with deviations in nearby cells. One possibility is to try larger quadrat sizes (e.g., see Greig-Smith 1964). If data are provided by county, such aggregation into arbitrarily larger quadrats or regions is not possible. Another possibility is to account explicitly for the spatial pattern in the deviations between observed and expected counts. This is the intent of the spatial chi-square statistic, discussed in Section 3.8.

3.3.5   Testing with Multiple Quadrat Sizes

It is natural to search over multiple quadrat sizes, to discern whether pattern exists at some scales and not others. If one uses the quadrat test at multiple scales, the probability of rejecting a true null hypothesis (i.e., a Type I error) will no longer be equal to α (a significance level associated with individual test); by chance alone, the analyst will be more likely than that to find a "significant" pattern in random data simply because multiple tests are carried out.

To illustrate, $n = 40$ points were located in a square, and $w = 5$ separate tests were carried out, on grids ranging from 2×2 to 6×6. This was repeated 1000 times; 21.8% of the time at least one of the five χ^2 values was significant when using $\alpha = 0.05$. Type I error probabilities for other values of n and w are shown in the last column of Table 3.3.

To retain an overall Type I error probability of α, it appears reasonable to derive a critical value of the test statistic by dividing α by the number of tests carried out (i.e., by applying a Bonferroni adjustment; see Chapter 5, Section 5.2). Supporting evidence comes from the following numerical experiment. n points were randomly scattered in a square, and w different grid sizes were used as overlays on the study area. The coarsest grid was 2×2, and w was chosen for each n to yield a finest grid that had on the order of one point per quadrat. There were 10,000 repetitions for each n; for

TABLE 3.3

Quadrat Analysis at Multiple Scales

n	w	5th Percentile of Minimum p	Bonferroni Adjustment (.05/w)	Type I Error Probability w/o Adjustment
20	3	.0186	.0167	.112
30	4	.0129	.0125	.194
40	5	.0098	.0100	.218
50	6	.0077	.0083	.254
80	8	.0052	.0062	.338

each repetition, the minimum p-value associated with the w χ^2 values was noted. The 500th smallest of the 10,000 minimum p-values was also noted (corresponding to $\alpha = 500/10,000 = 0.05$). The fifth percentiles of the minimum p-values are given in Table 3.3 for various values of n and w. Note the close correspondence between this column and the Bonferoni adjustment ($.05/w$). This implies that the best result across scales (i.e., the lowest p-value) should be compared with .05 (or more generally, α), divided by the number of tests (w). When min $p < \alpha/w$, the null hypothesis is rejected.

3.3.6 Optimal Quadrat Size: Appropriate Spatial Scales for Cluster Detection

One important issue in the use of quadrat analysis is the size of the quadrat; if the cell size is too small, there will be many empty cells, and if clustering exists on all but the smallest spatial scales, it will be missed. If the cell size is too large, one may miss patterns that occur within cells. One may find patterns on some spatial scales and not at others, and thus, the choice of quadrat size can seriously influence the results. Curtiss and McIntosh (1950) suggest an "optimal" quadrat size of two points per quadrat. Bailey and Gatrell (1995) suggest that the mean number of points per quadrat should be about 1.6.

However, to maximize the probability of finding a cluster of points, the quadrat size should match the size of the cluster; in addition, the cluster should be essentially contained within one of the quadrats, and not, for example, be half in one quadrat and half in an adjacent quadrat. Of course, this is generally difficult to achieve because the size and location of potential clusters is not generally known before the analysis is carried out!

To illustrate, 80 points were located randomly in the unit square (a square with x and y coordinates ranging from 0 to 1) with probability .9; with probability $p = .1$, they were located within the central area bounded by x- and y-coordinates of {0.4, 0.6}. A 5×5 grid was laid over the study area, and the quadrat statistic led to the rejection of the null hypothesis of randomness 59.7% of the time in 10,000 repetitions. Note that the middle cell of this 5×5 grid coincides precisely with the area of increased risk. Table 3.4 shows the results for other grid sizes. Note that the drop in power is quite sharp; in addition, the 2×2, 4×4, and 6×6 grids do not easily pick up the cluster because in these cases the cluster is split up among several grid cells.

This general principle of choosing a spatial scale that matches the hypothesized size of potential clusters is important in all forms of cluster detection, and we will return to it in Chapter 5. More generally, it is important to recognize that modifying not only the size but also the shape and orientation of the spatial units will have an effect on the results. This issue is known as the *modifiable areal unit problem* (MAUP), and it has a long history of investigation in many fields. Well-cited references in geography include

TABLE 3.4

Power of Global χ^2 Test for
Alternative Grid Sizes

Grid	Power
2×2	0.053
3×3	0.356
4×4	0.117
5×5	0.597
6×6	0.191
7×7	0.283
8×8	0.275
9×9	0.227
10×10	0.338

Openshaw (1984) and Fotheringham and Wong (1991); a more recent discussion in the context of health studies is provided by Gregorio et al. (2005).

3.3.7   A Comparison of Alternative Quadrat-Based Global Statistics

In this section, we compare the statistical power of four global statistics. One of these is the standard Pearson chi-square goodness-of-fit statistic. The second is based simply on the squared deviation between observations and expectations. The remaining two arise from transformations that have been used to convert regional observations and expectations into standard normal variables. It should be noted that this comparison is made here simply to illustrate the point that different statistics can have different power against different alternatives. A more complete and comprehensive discussion of the power of alternative goodness-of-fit tests is given by Cressie and Read (1984). The four global statistics are

1. The usual Pearson chi-square goodness-of-fit statistic.

2. A global statistic based on the sum of squared deviations between observed and expected values.

3. A global statistic based on the sum of squared z-scores, where each z-score is derived from Rossi's transformation (Rossi et al. 1999): $\frac{x-3e+2\sqrt{xe}}{2\sqrt{e}}$, where x and e represent the observed and expected counts, respectively. This transformation is an average of two alternative normalizing transformations, and it has been shown to have particularly good properties (such as little skewness in the resulting distribution), relative to other transformations.

4. A global statistic based on the sum of squared z-scores, where each z-score is derived from a square root transformation (Freeman and Tukey, 1950): $\sqrt{x} + \sqrt{x+1} - \sqrt{4e+1}$.

The comparison is carried out by Monte Carlo simulation of the following scenario.

There are five regions. The probability that a case falls into each of the five regions, under the null hypothesis, is taken to be {0.1, 0.1, 0.2, 0.3, 0.3}.

The null hypothesis was simulated 100,000 times, allocating $n = 50$ cases to the five regions. For each simulation, the expected number of cases falling in each region is therefore {5, 5, 10, 15, 15}. The 95th percentile of each statistic was found from these simulations as follows:

Pearson chi-square: 9.3667

Sum of squared deviations of observed and expected values: 94

Global statistic based on Rossi's transformation: 9.92

Global statistic based on square root transformation: 9.51

Next, several alternative hypotheses were examined. These were constructed by adding 0.08 to the null probability in a single region (creating raised risk in that region), and 0.02 was subtracted from the null probabilities associated with each of the other four regions. This was done for regions 1, 3, and 5, and thus:

H_{11}: $p = \{.18\ .08\ .18\ .28\ .28\}$

H_{13}: $p = \{.08\ .08\ .28\ .28\ .28\}$

H_{15}: $p = \{.08\ .08\ .18\ .28\ .38\}$

Results are shown in Table 3.5, which gives the statistical power for each combination of statistic and alternative hypothesis.

Entries in the table give the percentage of null hypotheses rejected, using the critical values derived earlier. Note that the chi-square test is best when the raised incidence occurs in a region with relatively low probability of receiving a case (H_{11}), whereas the sum of squared deviation statistic is best when there is excess risk in a region with a relatively high probability of receiving a case (H_{15}). These results are similar to those

TABLE 3.5

Probability of Correctly Rejecting Alternative False Hypotheses for Various Global Statistics

	H_{11}	H_{13}	H_{15}	Total
Chi-square	30.1	16.8	12.2	59.1
Sum of squared deviation	15.7	15.9	15.3	46.9
Rossi transformation	25.2	16.5	13.6	55.3
Square root transformation	21.8	16.0	13.7	51.5

found in Rogerson (1999) for spatial versions of these statistics. When the deviation occurs in a region with an average probability of receiving a case (H_{13}), all three statistics demonstrate roughly equal power.

Although the Rossi transformation is not best in any of the three, its overall performance is not bad, and it also performs well in the context of monitoring (see Chapter 8).

3.4 Spatial Dependence: Moran's *I*

Sometimes, point locations are not available, and data are given for areas only.

Moran's *I* statistic (1948) is one of the classic ways of measuring the degree of pattern (or spatial autocorrelation) in areal data. Moran's *I* is calculated as follows:

$$I = \frac{m \sum_i^m \sum_j^m w_{ij}(y_i - \bar{y})(y_j - \bar{y})}{\left(\sum_i^m \sum_j^m w_{ij}\right) \sum_i^m (y_i - \bar{y})^2} \tag{3.10}$$

where there are n regions, y_i is the observed value of the variable of interest in region i, $\bar{y}$ is the mean of y_i ($i = 1, \ldots, m$), and w_{ij} is a measure of the spatial proximity between regions i and j. It is interpreted much like a correlation coefficient. Values near +1 indicate a strong spatial pattern (high values tend to be located near one another, and low values tend to be located near one another). Values near −1 indicate strong negative spatial autocorrelation; high values tend to be located near low values. (Spatial patterns with negative autocorrelation are extremely rare; a notable exception is discussed in Griffith (2006)). Finally, values near zero indicate an absence of spatial pattern. It should be noted, however, that the range of Moran's *I* statistic does not exactly fit the interval [−1, 1], different from the ordinary correlation coefficient.

In addition to this descriptive interpretation, there is a statistical framework that allows one to decide whether any given pattern deviates significantly from a random pattern. One approximate test of significance is to assume that, when the number of regions n is sufficiently large (greater than about 20 or 30), *I* has a normal distribution with mean and variance equal to

$$E[I] = \frac{-1}{m-1}$$

$$V[I] = \frac{m^2(m-1)S_1 - m(m-1)S_2 + 2(m-2)S_0^2}{(m+1)(m-1)^2 S_0^2} \tag{3.11}$$

where

$$S_0 = \sum_i^m \sum_{j \neq i}^m w_{ij}$$

$$S_1 = 0.5 \sum_i^m \sum_{j \neq i}^m (w_{ij} + w_{ji})^2 \tag{3.12}$$

$$S_2 = \sum_k^m \left(\sum_j^m w_{kj} + \sum_i^m w_{ik} \right)^2$$

A small but growing number of software packages calculate the coefficient and its significance (e.g., GeoDa and the ArcGIS' Spatial Statistics toolbox).

The use of the normal distribution to test the null hypothesis of randomness relies on one of two assumptions:

1. Randomization: each permutation (rearrangement) of the observed regional values is equally likely.
2. Normality: the observed values are taken as arising from normal distributions having identical means and variances.

The formulae given above (Equations 3.11 and 3.12) for the variance assume that the normality assumption holds. The variance formula for the randomization assumption is algebraically more complex, and gives values that are only slightly different (see, e.g., Griffith 1987).

For the case of binary connectivity on a square grid, where adjacency is defined as "rook's case" (i.e., the weights are equal to one for the four adjacent regions sharing a side, and zero for all other regions), the variance expression may be approximated by $1/(2m)$. For the case of a hexagonal grid, or a "typical" regional configuration, the variance may be approximated by $1/(3m)$.

3.4.1 Illustration

Let us consider a simple example shown in Figure 3.2, consisting of 6 regions. The following is a binary connectivity weight matrix for the 6 regions, where $w_{ij} = 1$ if the regions i and j are adjacent, and $w_{ij} = 0$ otherwise. Note that $w_{ii} = 0$.

```
0 1 1 1 0 0
1 0 0 1 1 0
1 0 0 1 0 0
1 1 1 0 1 1
0 1 0 1 0 1
0 0 0 1 1 0
```

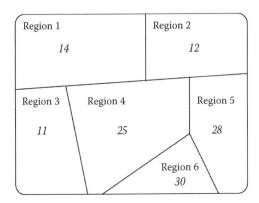

FIGURE 3.2
Sample problem for Moran's *I* and Geary's *C* analyses.

For this example, $m = 6$, $\bar{y} = \Sigma_i y_i/6 = 20$, $\Sigma_i \Sigma_j w_{ij} = 18$, and $\Sigma_i(y_i - \bar{y})^2 = 370$. The double summation part of the denominator in Equation 3.10 is calculated as

$$\sum_i \sum_j w_{ij}(y_i - \bar{y})(y_j - \bar{y}) = (14-20)(12-20) + (14-20)(11-20)$$

$$+ (14-20)(25-20) + (12-20)(14-20)$$

$$+ (12-20)(11-20) + (12-20)(28-20)$$

$$+ (11-20)(14-20) + (11-20)(25-20)$$

$$+ (25-20)(14-20) + (25-20)(12-20)$$

$$+ (25-20)(11-20) + (25-20)(28-20)$$

$$+ (25-20)(30-20) + (28-20)(12-20)$$

$$+ (25-20)(12-20) + (30-20)(12-20)$$

$$+ (30-20)(25-20) + (30-20)(28-20)$$

$$= 156$$

Therefore, Moran's *I* value is

$$I = \frac{m\sum_i^m \sum_j^m (y_i - \bar{y})(y_j - \bar{y})}{\left(\sum_i^m \sum_j^m w_{ij}\right)\sum_i^m (y_i - \bar{y})^2} = \frac{6 \cdot 186}{18 \cdot 370} = 0.1676$$

which indicates some positive spatial autocorrelation, although it is not strong.

3.4.2   Example: Low Birthweight Cases in California

The California low birthweight data described in Chapter 1, subsection 1.7.2.6 is first aggregated by county and then converted into rates as the number of LBW cases divided by the total number of births. Moran's *I* statistic applied to this rate data yields a value of $I = 0.195$, when rook's binary connectivity matrix is used. The expected value and variance are −0.018 and 0.007, respectively, and therefore, the z-score is 2.579, indicating significant positive spatial autocorrelation in the distribution of LBW rate at the 5% significance level.

When the rook's weight matrix is adjusted so that elements in each row sum to 1, the *I* value is 0.202, and the variance is 0.008 (note that the expected value remains the same regardless of the definition of the weight matrix). The associated z-score is 2.493, which is slightly less significant than the preceding result. This row adjustment of weight matrices is generally recommended to avoid assigning disproportionately large emphasis to regions with a large number of neighbors.

3.5   Geary's C

Geary's *C* statistic (1954), also called Geary's contiguity ratio, is another widely used method to measure the degree of spatial autocorrelation. It is calculated as

$$C = \frac{(m-1)\sum_{i=1}^{m}\sum_{j=1}^{m} w_{ij}(y_i - y_j)^2}{2\left(\sum_{i=1}^{m}\sum_{j\neq i} w_{ij}\right)\sum_{i=1}^{m}(y_i - \bar{y})^2} \tag{3.13}$$

and its value ranges approximately between 0 and 2. When positive spatial autocorrelation exists, pairs of regions that are close to each other tend to have similar values, so that Geary's *C* takes a value close to zero. When negative spatial autocorrelation exists, regions that are close to one another tend to have quite different values, so that Geary's *C* approaches the other extreme value of 2. No spatial autocorrelation would result in a value near 1.

The expected value of Geary's *C* statistic under the null hypothesis of no spatial pattern is 1, regardless of the number of areal units in a given study region; that is, $E(C) = 1$. The variance, under the normality assumption, is (Griffith 1987)

$$\sigma^2 = \left\{ (m-1) \left[\sum_i^m \sum_j^m w_{ij} + \sum_i^m \left[\sum_j^m w_{ij} \left(\sum_j^m w_{ij} - 1 \right) \right] \right] \Big/ 2 \right\} - \frac{\left(\sum_i^m \sum_j^m w_{ij} \right)^2}{2}$$

$$\times \left\{ (m+1) \left(\sum_i^m \sum_j^m \frac{w_{ij}}{2} \right)^2 \right\} \tag{3.14}$$

3.5.1   Illustration

Here, we apply Geary's C to the same example used in the previous section with Moran's I. The double summation part of the numerator in Equation 3.13 is calculated as

$$\sum_{i=1}^m \sum_{j=1}^m w_{ij}(y_i - y_j)^2 = (14-12)^2 + (14-11)^2 + (14-25)^2 + (12-14)^2$$
$$+ (12-25)^2 + (12-28)^2 + (11-14)^2 + (11-25)^2$$
$$+ (25-14)^2 + (25-12)^2 + (25-11)^2 + (25-28)^2$$
$$+ (25-30)^2 + (28-12)^2 + (28-25)^2 + (28-30)^2$$
$$+ (30-25)^2 + (30-28)^2$$
$$= 1586$$

Geary's C value is thus obtained as

$$C = \frac{(m-1)\sum_{i=1}^m \sum_{j=1}^m w_{ij}(y_i - y_j)^2}{2\left(\sum_{i=1}^m \sum_{j \neq i}^m w_{ij}\right)\sum_{i=1}^m (y_i - \bar{y})^2} = \frac{(6-1)\cdot 1586}{2\cdot 18\cdot 370} = 0.5953$$

Because the obtained value is between 0 and 1, positive spatial autocorrelation in the example spatial pattern is suggested again.

3.5.2   Example: Low Birthweight Cases in California

Geary's C statistic applied to the LBW rate data used in subsection 3.4.2 yields a value of $C = 0.785$ when rook's binary connectivity matrix is used.

The variance computed by Equation 3.14 is 0.010, leading to the z-score of −2.127, which again indicates significantly positive spatial autocorrelation at the 5% significance level. When the row-adjusted version of the rook's weight matrix is used, the C value is 0.748 and the variance is 0.033. The associated z-score is −1.391, indicating still positive but insignificant spatial autocorrelation in the distribution of LBW rates in California.

3.6   A Comparison of Moran's *I* and Geary's *C*

Figure 3.3 highlights the differences between pairs of observations that tend to contribute to a significant *I*, and those that tend to contribute to a significant *C*. For simplicity, we assume a binary connectivity definition for the weights; $w_{ij} = 1$ if the regions *i* and *j* are adjacent, and $w_{ij} = 0$ otherwise. The horizontal and vertical axes represent values of the variable observed in regions *i* and *j*, respectively, and points plotted on the figure would represent observations for adjacent pairs of regions. Note that if all pairs of adjacent regions are plotted, two points will be plotted for each pair of regions, and they will be symmetric with respect to the 45° line ($y_i = y_j$). Points near the main diagonal (regions A and C) will contribute to a significant finding of positive spatial autocorrelation using Geary's *C*. Points far from the diagonal (regions D, E, and F) will contribute to a

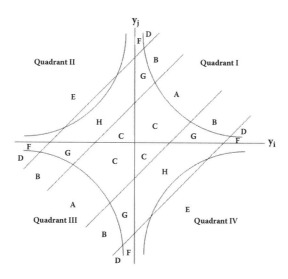

FIGURE 3.3
Contributions of adjacent point pairs to Moran and Geary statistics.

finding of significant negative spatial autocorrelation when Geary's C is used because the difference $(y_i - y_j)^2$ will be large. Intermediate distances from the diagonal (regions B, G, and H) will lead to a finding of no significant spatial autocorrelation, using C.

When both y_i and y_j are high (or when they are both low), this will contribute to positive spatial autocorrelation as measured by I. Such observations correspond to those falling in regions A, B, and D in the figure, and these regions are delineated by hyperbolic lines corresponding to equal values of $(y_i - \bar{y})(y_j - \bar{y})$, and where, without loss of generality, we have taken the mean to be zero (for variables with nonzero means, the mean can be subtracted from each observation to give a new, transformed observation with zero mean). Observations on the other side of these hyperbolic lines, in regions C, G, H, and F, contribute to a finding of no significant spatial autocorrelation, using I. Observations in the second and fourth quadrants, where one observation is far below the mean and the other observation is far above the mean (region E), will contribute to significant negative spatial autocorrelation when I is used.

3.6.1   Example: Spatial Variation in Handedness in the United States

The second National Health and Nutrition Examination Survey, NHANES II, covers the period 1976–1980, and is a nationwide probability sample of 27,801 persons from 6 months to 74 years of age. From this sample, 25,286 people were interviewed, and 20,322 people were examined. There were 17,386 born in the United States and answering the question on handedness; 1,493 were left-handed (8.59%). Details on NHANES II are available at http://www.cdc.gov/nchs/about/major/nhanes/nhanesii.htm.

Because of small sample sizes in some states, 10 western states were pooled together (Arizona, Colorado, Montana, Nebraska, Nevada, New Mexico, North Dakota, South Dakota, Utah, and Wyoming). In addition, Delaware, Maryland, New Hampshire, Rhode Island, Vermont, and West Virginia were not included, again because of small sample sizes. Figure 3.4 depicts how the proportion of the population that is left-handed varies by state.

A binary connectivity adjacency matrix was constructed by setting the weight equal to one when pairs of states shared a common boundary, and was set equal to zero otherwise. Moran coefficients and Geary ratios were computed for both white and nonwhite populations; the results are shown in Table 3.6.

The results show that, for the total population, both measures of spatial autocorrelation are significant, with p values less than .05. For the white population, the Geary ratio is significant, whereas the Moran coefficient is not. Interestingly, the opposite is true for the nonwhite population—the Moran coefficient is significant, whereas the Geary ratio is not.

Figure 3.5 displays a plot of pairs of adjacent states for the white population. The figure is symmetric about the 45° line because all pairs of points

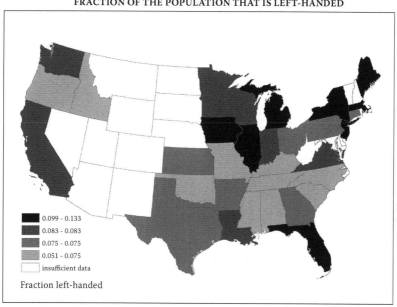

FRACTION OF THE POPULATION THAT IS LEFT-HANDED

■	0.099 - 0.133
▓	0.083 - 0.083
▒	0.075 - 0.075
░	0.051 - 0.075
□	insufficient data

Fraction left-handed

FIGURE 3.4
Geographic variation in handedness in the United States.

are plotted (i.e., the region $i - j$ pair is plotted twice). In Figure 3.5, there are many points close to the 45° line (corresponding to regions A and C in Figure 3.3). The corresponding figure for the nonwhite population (not shown) has points both close to and far from the 45° line, and hence the Geary ratio is not significant for the nonwhite population. In Figure 3.5, there are few points in quadrants I and III that are far from the origin. Hence, there are no strong contributions to Moran's I, and it is insignificant. The corresponding figure for the nonwhite population does have points in this region (corresponding to regions B and D in Figure 3.3), and hence Moran's I is significant for the nonwhite population.

The nature of Figure 3.3 and the location of the points in Figure 3.5 lead to the question of whether there are variables that are characterized by

TABLE 3.6

Moran Coefficients and Geary Ratios for Handedness Data

	Total	White	Nonwhite
Moran	0.173	0.090	0.302
	($p = .048$)	($p = .16$)	($p < .02$)
Geary	0.597	0.418	0.645
	($p < .01$)	($p < .01$)	($p = .11$)

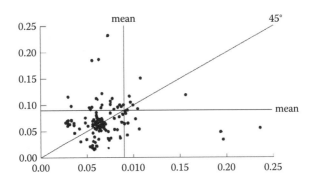

FIGURE 3.5
Contributions of adjacent state pairs to spatial autocorrelation of handedness: white population.

certain types of statistical distribution and spatial processes that might naturally lead to situations where either Moran's *I* or Geary's *C* is to be preferred in detecting spatial autocorrelation. We explore this further in subsection 3.6.2.

3.6.2 Statistical Power of *I* and *C*

Moran's *I* is generally regarded as having higher statistical power in comparison with Geary's *C*; therefore, it is better able to detect spatial autocorrelation when it exists (Cliff and Ord 1981, Haining 1990). In this section, we confirm that this is the case for common types of spatial processes when the regional values have a normal distribution. In addition, we compare the power of *I* and *C* when the regional values have distributions other than normal.

Positive spatial autocorrelation was induced into a 5 × 5 lattice, using a spatial autoregressive (SAR) model and the simulation procedure described by Haining (1990). Four outcomes are possible: (a) both *C* and *I* are significant; (b) both *C* and *I* are insignificant; (c) *C* is significant, and *I* is not; and (d) *I* is significant, and *C* is not. Table 3.7 displays examples of the latter two categories. In panels (a) and (b) of Table 3.7, there is a relatively large number of pairs of adjacent regions where one value is slightly above the mean and the other is slightly below; these lead to significant Geary ratios but insignificant Moran coefficients. In panels (c) and (d), there are relatively few such pairs.

Table 3.8 demonstrates that when normally distributed variables are used to generate observations that follow a spatial autoregressive model, Moran's *I* is more powerful than Geary's *C* in detecting the spatial autocorrelation. The table is based on the results of 1000 simulations for each level, ρ, of spatial autocorrelation; the last four elements in each row sum to 1 and give the proportion of all simulations that fall into each of the four

TABLE 3.7

Spatial Distributions of Variables in Special Cases Where One of the Two
Measures of Spatial Autocorrelation Is Significant

(a) C, Significant; I, Insignificant

1.04	1.75	0.26	−0.73	−2.31
−0.44	−0.68	−0.73	0.61	−1.17
1.42	−0.26	0.03	−0.96	−0.99
−0.81	0.76	0.85	1.65	0.68
0.08	0.50	1.08	0.12	1.13

(mean = 0.1152, standard deviation = 0.99)

(b) C, Significant; I, Insignificant

2.73	1.08	2.18	0.09	−0.26
1.69	0.74	−0.78	−0.28	0.11
0.04	−0.28	−0.84	1.41	−1.24
0.65	0.30	−0.49	−0.63	−0.08
0.18	1.18	−1.98	−0.98	1.20

(mean = 0.2296, standard deviation = 1.09)

(c) I, Significant; C, Insignificant

1.31	1.46	1.76	2.36	0.88
2.05	0.76	1.38	2.34	−0.75
1.29	−1.84	−0.15	−1.40	−0.08
0.20	−0.73	0.75	−0.42	0.93
−0.64	0.13	3.08	0.03	0.49

(mean = 0.6076, standard deviation = 1.20)

(d) I, Significant; C, Insignificant

−0.64	−0.54	1.60	1.40	0.11
0.29	0.99	1.82	−1.06	0.11
−1.07	−1.05	0.61	−1.31	0.26
−0.68	−0.41	−0.33	−0.82	−0.74
−0.32	0.56	−0.38	−0.49	−0.24

(mean = −0.09, standard deviation = 0.84)

TABLE 3.8

SAR Model with Normally Distributed Variables on a 5×5 Lattice and Rook's
Case Adjacency

ρ	Power I	C	I Is Significant C Is Insignificant	C Is Significant I Is Insignificant	Neither Is Significant	Both Are Significant
0.05	0.152	0.133	0.041	0.022	0.826	0.111
0.10	0.291	0.265	0.073	0.047	0.662	0.218
0.15	0.523	0.461	0.090	0.028	0.449	0.433
0.20	0.793	0.757	0.069	0.033	0.174	0.724
0.25	0.942	0.926	0.026	0.010	0.048	0.916

TABLE 3.9

MA Model with Normally Distributed Variables on a 5 × 5 Lattice and Rook's Case Adjacency

Power			I Is Significant C Is Insignificant	C Is Significant I Is Insignificant	Neither Is Significant	Both Are Significant
ρ	I	C				
0.05	0.131	0.130	0.037	0.036	0.833	0.094
0.10	0.287	0.262	0.059	0.044	0.669	0.218
0.15	0.485	0.438	0.082	0.035	0.480	0.403
0.20	0.676	0.633	0.086	0.043	0.281	0.590
0.25	0.810	0.735	0.102	0.027	0.163	0.708

categories describing the significance of the results. Note that although Moran's I is relatively more powerful, approximately 3% to 4% of the time Geary's C will be significant, whereas Moran's I will be insignificant.

Moving average (MA) models of spatial association were also simulated, and the results are quite similar to the spatial autoregressive model. Moran's I is again more powerful (see Table 3.9), although C will be found significant in 3% to 4% of the cases where I is insignificant.

We repeated these experiments using exponential and uniform distributions as well as a negatively skewed distribution created by subtracting an exponentially distributed variable from the maximum of a large number of simulations of the same exponential distribution. For both the exponential and negatively skewed distributions, results (not shown here) were similar to those described earlier, and the power of Moran's I was greater than Geary's C. Results for the uniform distribution are shown in Table 3.10. Here, Geary's C appears to have slightly higher relative power in the presence of small amounts of spatial autocorrelation. Note, though, that I is slightly more powerful than C when autocorrelation is more pronounced. This finding is relevant in assessing maps of statistical significance (e.g., maps of p values). Under the null hypothesis of no raised incidence in any region, p values have a uniform distribution on the (0,1) interval. Maps of p values under this null hypothesis should display no

TABLE 3.10

SAR Model with Uniformly Distributed Variables on a 5 × 5 Lattice and Rook's Case Adjacency

Power			I Is Significant C Is Insignificant	C Is Significant I Is Insignificant	Neither Is Significant	Both Are Significant
ρ	I	C				
0.05	0.123	0.143	0.023	0.043	0.832	0.100
0.10	0.309	0.313	0.040	0.044	0.647	0.269
0.15	0.775	0.740	0.072	0.037	0.188	0.703
0.20	0.996	0.994	0.004	0.002	0.002	0.992

spatial autocorrelation. It is of interest to assess such maps for the presence of spatial autocorrelation because this could point not only to spatial clusters of similar statistical significance but also to spatial clusters of significantly high (or low) incidence, depending on the particular *p* values. We conclude that for this situation, there is no particular advantage in using Moran's *I*, and in fact, there may be a slight advantage in using Geary's *C*, especially when the level of autocorrelation among the *p*-values is small.

This exercise has served to show that the two statistics can indeed yield different conclusions in applied settings. The simulation experiments have shown that Moran's *I* may have slightly better power in general, but that there are also situations (e.g., analyzing maps of *p*-values) in which Geary's *C* may be preferred.

More generally, the comparison between Moran's *I* and Geary's *C* serves to emphasize the point that there is no single "best" statistic. Moran's *I* will do better against some alternatives (in the sense of rejecting false null hypotheses more often), and Geary's *C* will do better against others. If the analyst is able to come up with a well-specified alternative hypothesis, this can aid in the choice of a statistic (namely, one with good power against that alternative). If not, calculation of several alternative statistics can be useful in pointing the analyst toward appropriate alternative hypotheses (namely, those alternatives for which the statistics with the lowest *p*-values were designed, or have good power).

3.7 Oden's I_{pop} Statistic

One of the characteristics of Moran's *I* is that within-region variations can undermine the validity of the randomization or normality assumptions. For example, regions with small populations may be expected to exhibit more variability. Oden (1995) accounts for this within-region variation explicitly by modifying *I* as follows:

$$I_{pop} = \frac{n^2 \sum_i^m \sum_j^m w_{ij}(r_i - p_i)(r_j - p_j) - n(1 - 2\bar{b})\sum_i^m w_{ii}r_i - n\bar{b}\sum_i^m w_{ii}p_i}{S_0 \bar{b}(1 - \bar{b})} \tag{3.15}$$

where r_i and p_i are the observed and expected proportions of all cases falling in region *i*, respectively. Furthermore, there are *m* regions, *n* incidents, and a total base population of *x*. The overall prevalence rate is $\bar{b} = n/x$. Also,

$$S_0 = x^2 A - xB \tag{3.16}$$

where

$$A = \sum_{i}^{m}\sum_{j}^{m} p_i p_j w_{ij}$$

$$(3.17)$$

$$B = \sum_{i} p_i w_{ii}$$

Oden suggests that statistical significance be evaluated via a normal distribution, with mean and variance

$$E[I_{pop}] = \frac{-1}{x-1}$$

$$(3.18)$$

$$V[I_{pop}] \approx \frac{2A^2 + C/2 - E}{A^2 x^2}$$

where A is defined as earlier, and

$$C = \sum_{i}^{m}\sum_{j}^{m} p_i p_j (w_{ij} + w_{ji})^2$$

$$(3.19)$$

$$E = \sum_{i}^{m} p_i \left[\sum_{j}^{m} p_j (w_{ij} + w_{ji}) \right]^2$$

3.7.1   Illustration

For the six-region study area shown in Figure 3.2, let us assume that the population of regions 1, 2, 3, and 6 is 100 each and that of regions 4 and 5 is 300 each. Under the assumption that the expected number of cases in each region is proportional to the population, r_i and p_i for the six regions are given in Table 3.11. In order to account for within-region variations, the binary connectivity matrix used earlier needs to be modified so that $w_{ii} \neq 0$; for simplicity, $w_{ii} = 1$ $(i = 1,\dots,6)$ is used in this example.

Here, $n = 120$ and $x = 1,000$ so that $\bar{b} = 120/1{,}000 = 0.12$. $A = 0.8$ and $B = 1.0$, which makes $S_0 = 799{,}000$ and $S_0 \bar{b}(1-\bar{b}) = 84{,}374.4$. The numerator of Equation 3.15 is obtained as

$$n^2 \sum_{i}^{m}\sum_{j}^{m} w_{ij}(r_i - p_i)(r_j - p_j) - n(1 - 2\bar{b})\sum_{i}^{m} w_{ii} r_i - n\bar{b}\sum_{i}^{m} w_{ii} p_i$$

$$(3.20)$$

$$= 120^2 (-0.00139) - 120(1 - 0.24)(1.0) - 120(0.12)(1.0) = -125.6$$

TABLE 3.11

Observed and Expected Proportions of Cases Falling in Region i (for the Zone System Shown in Figure 3.2)

Region i	Observed Cases	Population	Observed Proportions (r_i)	Expected Proportions (p_i)	Difference ($r_i - p_i$)
1	14	100	0.117	0.1	0.017
2	12	100	0.100	0.1	0.000
3	11	100	0.092	0.1	−0.008
4	25	300	0.208	0.3	−0.092
5	28	300	0.233	0.3	−0.067
6	30	100	0.250	0.1	0.150
Total	120	1000	1.000	1.0	0

Therefore, Oden's I_{pop} statistic is given by

$$I_{pop} = \frac{-125.6}{84374.4} = -0.00149 \tag{3.21}$$

and the associated expected value and variance are $E[I_{pop}] = \frac{-1}{1000-1} = -0.001$ and $V[I_{pop}] = 3.375e^{-7}$, respectively. The z-score of the observed I_{pop} value is −0.839, implying that the observed pattern does not have any significant pattern.

3.8 Tango's Statistic and a Spatial Chi-Square Statistic

Tango (1995) suggested the following global statistic to detect clusters:

$$C_G = \sum_i^m \sum_j^m w_{ij}(r_i - p_i)(r_j - p_j) \tag{3.22}$$

In matrix form,

$$C_G = (\mathbf{r} - \mathbf{p})'\mathbf{W}(\mathbf{r} - \mathbf{p}) \tag{3.23}$$

where $\mathbf{r}$ and $\mathbf{p}$ are $m \times 1$ vectors with elements containing the observed and expected proportions of cases in each region, and $\mathbf{W}$ is a matrix containing elements w_{ij} that measure the closeness (or connectivity) between regions i and j. To test the null hypothesis that the incidence pattern is random, Tango first gives the expected value and variance of the statistic as

$$E[C_G] = \frac{1}{N} Tr(\mathbf{W}\mathbf{V}_p)$$

(3.24)

$$V[C_G] = \frac{2}{N^2} Tr(\mathbf{W}\mathbf{V}_p)^2$$

where N is the total number of observed cases, $Tr(\cdot)$ is the matrix trace of a matrix defined as the sum of diagonal elements, and

$$\mathbf{V}_p = \Delta\mathbf{p} - \mathbf{p}\mathbf{p}'$$

(3.25)

with $\Delta\mathbf{p}$ defined as a $m \times m$ diagonal matrix containing the elements of $\mathbf{p}$ on the diagonal. Tango then finds that the test statistic

$$v + \frac{C_G - E[C_G]}{\sqrt{V[C_G]}}\sqrt{2v}$$

(3.26)

has an approximate chi-square distribution with v degrees of freedom, where

$$v = \left(\frac{Tr(\mathbf{W}\mathbf{V}_p^2)^{1.5}}{Tr(\mathbf{W}\mathbf{V}_p^3)}\right)^2$$

(3.27)

Tango's statistic is a weighted average of the covariations of deviations between observed and expected frequencies, for all pairs of points.

Rogerson (1999) developed and evaluated a spatial chi-square statistic that can be used as a global test of clustering. The statistic is defined as

$$R = \sum_i^m \sum_j^m \frac{w_{ij}(r_i - p_i)(r_j - p_j)}{\sqrt{p_i p_j}}$$

(3.28)

Note that this may be written as a combination of a chi-square goodness-of-fit statistic and a Moran-type statistic:

$$R = \sum_i^m \frac{w_{ii}(r_i - p_i)^2}{p_i} + \sum_i^m \sum_{j \neq i}^m \frac{w_{ij}(r_i - p_i)(r_j - p_j)}{\sqrt{p_i p_j}}$$

(3.29)

The statistic R will be large when either there are large deviations between observed and expected values within regions, or when nearby pairs of regions have similar deviations. Similar to Tango's statistic, R combines the features of quadrat analysis, which focuses on what goes on within cells, and Moran's I, which focuses on what the joint variation of pairs of nearby cells is. R is actually a special case of Tango's C_G, where Tango's weights are modified by dividing them by $\sqrt{p_i p_j}$. Thus, the distribution theory discussed for Tango's statistic may be adapted when using R to test the null hypothesis of randomness.

3.8.1 Illustration

For the simple example used in the Section 3.8, Tango's statistic is calculated as

$$C_G = \sum_i^m \sum_j^m w_{ij}(r_i - p_i)(r_j - p_j) = -0.00139 \tag{3.30}$$

and the associated expected value and variance are 0.00167 and 0.000015, respectively. The degrees of freedom obtained by Equation 3.27 is $v = 2.986$. Then, the test statistic for the observed C_G value is given by

$$v + \frac{C_G - E[C_G]}{\sqrt{V[C_G]}}\sqrt{2v} = 2.986 + \frac{-0.00139 - 0.00167}{\sqrt{0.000015}}\sqrt{2 \cdot 2.986} \approx 1.046 \tag{3.31}$$

This test statistic is considered to approximately follow the χ^2 distribution, and it is not significant (implying no significant spatial pattern) when compared with the critical value of the distribution (based on a Type I error probability of .05, and 2.986 degrees of freedom).

3.8.2 Example: Sudden Infant Death Syndrome in North Carolina

As seen in subsection 3.3.2.2, the SIDS data for 1974 deviate from spatial randomness in the sense that there is a significant "global" discrepancy between observed and expected county frequencies. Might such discrepancies remain at a larger geographic scale?

Using *GeoSurveillance* with the SIDS data (as described in subsection 3.3.2.2) choose Unadjusted under Variables (corresponding to Tango's statistic, which focuses on the difference between observed and expected for pairs of regions), and under "A range of values" in Bandwidth, choose from 0.2 to 2.0 by 0.2, which are the default values. This setting is to examine various specifications for w_{ij} (details of which will be explained) in Equation 3.22. Click Run.

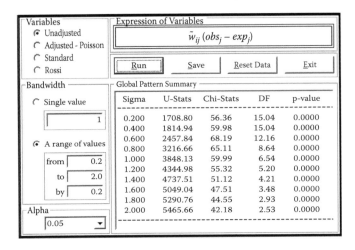

FIGURE 3.6
GeoSurveillance output for the application of Tango's C_G statistic to the North Carolina SIDS data.

Tango's statistic is calculated for the range of σ values as shown in Figure 3.6. Note that the highest test statistic under the Chi-stats column occurs for σ = 0.6, which implies that some spatial effects are important (i.e., σ > 0), and thus, the most significant clustering appears to occur on a scale greater than that of the individual county.

Rerunning this example using the Adjusted (spatial chi-square) option again reveals that results are significant across a range of spatial scales; in this case, the peak that occurred in the previous example for σ = 0.6 is absent. This combination of results would be expected when the cluster contains a region with a relatively large number of cases. When clusters are characterized by a collection of regions that have small expected numbers, the opposite results are likely to occur (where the spatial chi-square test [i.e., Adjusted] shows a peak at nonzero σ, and the unadjusted, Tango-type test, does not).

This approach of searching over various σ values should be considered exploratory in nature. If one tests multiple hypotheses (in this case using multiple σ-values), it is important to account for the multiple testing; otherwise significance would be expected if one simply tested enough hypotheses. Tango (2000) describes how multiple testing can be taken into account with his test by first searching over a range of σ-values for the best one, where "best" is defined as that value which is associated with the lowest p-value. Then this p-value is compared with the result of many minimal p-values found via Monte Carlo simulation of the null hypothesis over this same range of σ values. If the observed *p*-value is very low relative to these simulated values, then the null hypothesis is rejected and σ gives an estimate of cluster size.

3.9 Getis and Ord's Global Statistic

The *G(d)* statistic that Getis and Ord (1992) proposed is another well-known measure of spatial association. It is denoted as follows:

$$G(d) = \frac{\sum_{i=1}^{m} \sum_{j=1}^{m} w_{ij}(d) y_i y_j}{\sum_{i=1}^{m} \sum_{j=1}^{m} y_i y_j}, \quad i \neq j \tag{3.32}$$

where *m* is the number of regions and $w_{ij}(d)$ is the weight based on the spatial proximity between regions *i* and *j*. This statistic measures spatial association, using all locational pairs of values (y_i, y_j) such that location *i* and *j* are within distance *d* of each other. Similar to the global *I* statistic, it is common to use the standardized *G* statistic (formed by subtracting the expectation and then dividing the result by the standard deviation; see Ord and Getis (1995) for details), and then assuming normality to test the null hypothesis of no clustering. For a dominant pattern of high values near other high values, the *G* statistic is high, along with a high positive *z*-value; it is low when there is an overall tendency of clustering of low values. Getis and Ord provide the mean, variance, and *z*-statistic as follows:

$$Z[G(d)] = \frac{G(d) - E[G(d)]}{\sqrt{Var[G(d)]}} \tag{3.33}$$

$$Var[G(d)] = E[G^2(d)] - E[G(d)], \tag{3.34}$$

$$E[G(d)] = \frac{W}{m(m-1)}, \quad W = \sum_{i=1}^{m} \sum_{j=1}^{m} w_{ij}(d), \quad i \neq j, \tag{3.35}$$

$$E[G^2(d)] = \frac{1}{\left(m_1^2 - m_2\right)^2 n^{(4)}} \left[B_0 m_2^2 + B_1 m_4 + B_2 m_1^2 m_2 + B_3 m_1 m_3 + B_4 m_1^4 \right], \tag{3.36}$$

where

$$B_0 = (m^2 - 3m + 3)S_1 - mS_2 + 3W^2; \qquad B_1 = -[(m^2 - m)S_1 - 2mS_2 + 6W^2];$$

$$B_2 = -[2mS_1 - (m+3)S_2 + 6W^2]; \qquad B_3 = 4(m-1)S_1 - 2(m+1)S_2 + 8W^2;$$

$$B_4 = S_1 - S_2 + W^2;$$

$$S_1 = \frac{1}{2}\sum_i\sum_j (w_{ij} + w_{ji})^2, \ j \neq i; \qquad S_2 = \sum_i (w_{i.} + w_{.i}); \ w_{i.} = \sum_j w_{ij}, \ j \neq i;$$

$$m_j = \sum_{i=1} y_i^j, \ j = 1, 2, 3, 4; \qquad m^{(r)} = m(m-1)(m-2) \ (m-r+1).$$

$$(3.37)$$

In empirical examples, Getis and Ord emphasize the conjunctional use of the *I* statistic with *G*. If either a spatial pattern of high or low values is dominant, the global *I* statistic will show a highly positive value. The *G* statistic is able to discriminate between the two patterns. In a pattern where clustering of high values is dominant, the *G* statistic will be high; it will be low if low values cluster. Although the global *I* statistic is a measure of spatial autocorrelation, the *G* statistic can be seen as a measure of spatial concentration. In their case study using house prices data by ZIP code in San Diego, the standardized *G(d)* statistics showed negative values over the range of distance because there was a strong tendency for the low housing prices to be clustered around the central city. Subregions with high price values along the coast have fewer near neighbors with similar values than do the central city subregions. In contrast, the *Z[I]*s were all positive (Getis and Ord 1992).

3.9.1   Example: Low Birthweight Cases in California

The LBW rate data used in subsections 3.4.2 and 3.5.2 is again examined by $G^*(d)$, which is a variant of *G(d)* statistic, where $w_{ii}(d)$ is set to be non-zero and the condition of $i \neq j$ is removed from Equation 3.32. Although the original definitions of the *G(d)* and $G^*(d)$ statistics use a distance-based weight matrix $w_{ij}(d)$, other definitions of the weight matrix are also allowed. Therefore, for comparison purposes, the same binary weight matrix based on rook's connectivity used in the previous examples is applied here, too. The $G^*(d)$ statistic yields a value of $G^* = 0.081$, and the expected value and variance are 0.080 and 0.000004, respectively. The associated *z*-score is calculated as 0.459, indicating no significant clustering pattern in the distribution. This result might seem somewhat counterintuitive, considering the findings by Moran's I and Geary's C. However, because the G statistic attempts to capture clustering of larger-than-average values and that of lower-than-values simultaneously, it could be insignificant if both clustering tendencies exist in the data and cancel out each other.

3.10   Case–Control Data: The Cuzick–Edwards Test

The Cuzick–Edwards test (Cuzick and Edwards 1990) is a global or general test for clustering, for use with case–control data. The test statistic is simply a count of the k-nearest neighbors of a case that are also cases, summed over all cases. More formally,

$$T_k = \sum_i \sum_j w_{ij} \delta_i \delta_j \tag{3.38}$$

where δ_i is equal to one if location i is a case, and is equal to zero if location i is a control. The term w_{ij} is equal to one if j is a k-nearest neighbor of i, and zero otherwise. The choice of k determines the spatial scale at which the test is carried out.

Under the null hypothesis that the set of all cases and controls has been assigned to the set of all locations randomly (also known as *random labeling*), the expected value of T_k is

$$E[T_k] = \frac{kn_0(n_0 - 1)}{n - 1} \tag{3.39}$$

where n_0 is the number of cases and $n = n_0 + n_1$ is the number of cases and controls.

The variance of T_k is

$$V[T_k] = (kn + N_s)p_1(1 - p_1) + \{(3k^2 - k)n + N_t - 2N_s\}\left(p_2 - p_1^2\right)$$
$$- \{k^2(n^2 - 3n) + N_s - N_t\}\left(p_1^2 - p_3\right) \tag{3.40}$$

where

$$p_j = \prod_{i=0}^{j} \frac{n_0 - i}{n - i} \tag{3.41}$$

Furthermore, N_s is twice the number of pairs of points that are k-nearest neighbors of each other:

$$N_s = \sum_i \sum_j w_{ij} w_{ji} \qquad (3.42)$$

In addition,

$$N_t = \sum_j \sum_i \sum_{l \neq i} w_{ij} w_{lj} = \sum_j \left(\sum_i w_{ij} \right) \left\{ \left(\sum_i w_{ij} \right) - 1 \right\} \qquad (3.43)$$

The quantity $z = \{T_k - E[T_k]\}/\sqrt{V[T_k]}$ may be treated as a standard normal variable, and its significance assessed in the usual way.

3.10.1   Illustration

Using the data from Chapter 2, Table 2.1 (repeated in Table 3.1), and treating the first five points as cases, and the next five as controls, the value of T_1 is 4, implying that there are four cases that have cases as their nearest neighbor. The expected value of the statistic is $5(4)/9 = 2.22$. Using the formula for the variance, we find first that $p_1 = 2/9$, $p_2 = 1/12$, and $p_3 = 1/42$. Also, $N_s = 8$ and $N_t = 2$. Thus $V[T_1] = 1.37$, implying a z-score of $(4 - 2.22)/\sqrt{1.37} = 1.52$. This does not exceed the usual critical value of 1.645 that is used for a one-sided test with $\alpha = 0.05$, implying that we should not reject the null hypothesis, and we, therefore, conclude that the cases are not clustered. Caution should be used in the interpretation because the sample size is small; the normality of the z-score is asymptotic, and is therefore typically achieved only in larger samples.

3.11   A Global Quadrat Test of Clustering for Case–Control Data

Let C be the number of cases, and N be the number of controls. We wish to test the null hypothesis that the pattern of cases does not differ from what is expected, where "expected" is defined here on the basis of the control locations, which are assumed fixed. The primary impetus and rationale for the following test is the desire for a quick and simple test of the null hypothesis of no spatial clustering of cases relative to controls.

We begin by letting each control location be the center of a Thiessen polygon; thus, locations inside the polygon are closer to its center than to any other center. The null hypothesis is that the probability that a case falls into any particular polygon is $1/N$.

Now let C_i be the number of cases that are inside polygon i; these cases are closer to control i than to any other control. Conditional on the total number of cases C, the null hypothesis is that the cases are distributed across polygons according to a multinomial distribution. (Note that the null hypothesis is not that of random labeling of cases and controls.) Furthermore, each polygon is assumed to have an expectation of C/N cases under the null hypothesis. We then use the chi-square test statistic comparing observed and expected cases in each polygon:

$$\chi^2 = \sum_{i=1}^{N} \frac{(C_i - C/N)^2}{C/N} = \frac{N}{C} \sum_i (C_i - C/N)^2 = \left\{ \frac{N}{C} \sum_i^N C_i^2 \right\} - C \quad (3.44)$$

This will have a chi-square distribution with $N - 1$ degrees of freedom under the null hypothesis, conditional on the number of cases. Following the guidelines of Koehler and Larntz, the chi-square approximation for the null distribution will be realistic when $N \geq 3$ and $C^2/N > 10$. Furthermore, expectations should be greater than about $C/N = 0.25$.

We will be interested in high values of the statistic caused by a high number of cases in a subset of the polygons formed around the controls. Note that in the special and not uncommon case of $C = N$,

$$\chi^2 = \sum_i (C_i - 1)^2 = \left\{ \sum_i C_i^2 \right\} - C \quad (3.45)$$

The simplicity of the approach derives from the fact that the expected number of cases is the same in each polygon; implementation is possible because requirements for the magnitude of the expectation in each polygon are not as stringent as is commonly thought. A potential limitation of the approach is the requirement that cases be centered around controls instead of cases. This requirement could result in a cluster of cases being split up evenly, with individual cases in the cluster assigned to different controls, thereby weakening the power of the statistic to detect the cluster.

If the global statistic is significant, or even if it is not, it will often be of interest to look at the local statistics

$$z_i = \frac{(C_i - 0.5 - C/N)}{\sqrt{C/N}} \quad (3.46)$$

where the local statistics include a correction for continuity (although these z-scores will not necessarily be accurate when the expected number of cases in a cell is small).

It is also possible to construct a test by taking the C case locations as fixed, and then examining the distribution of the N controls across the C Thiessen polygons generated by the case locations. Under the null hypothesis, the spatial distribution of controls should be multinomial, with a constant expectation N/C in each of the polygons. This approach adopts the perspective that the case locations are fixed, and interest lies in evaluating how unusual the selection of control locations is, under the null hypothesis of no difference in the spatial distributions of cases and controls.

Letting N_i be the number of controls that are closer to case i than to any other case, the usual chi-square statistic associated with differences between observed and expected counts (of, in this instance, controls) is

$$\chi^2 = \sum_{i=1}^{C} \frac{(N_i - N/C)^2}{N/C} = \frac{C}{N} \sum_i (N_i - N/C)^2 \qquad (3.47)$$

This perspective may be useful in identifying "cool" spots, where there are significantly more controls near a particular case than expected.

3.11.1 Example

For the leukemia and lymphoma data described by Cuzick and Edwards (1990), there are 62 cases and 141 controls. Figure 3.7 displays the locations

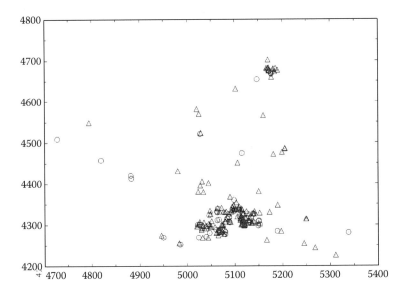

FIGURE 3.7
Childhood leukemia and lymphoma cases and controls in the north Humberside region of England, 1974–1986 (Cuzich and Edwards, 1990).

of cases (represented by circles) and controls (represented by triangles). We have $N = 141$ regions around the controls, each with the expectation of 62/141 cases. For each control (i), we count the number of cases that are closer to that control than to any other control. Call this count C_i. The χ^2 statistic is

$$\chi^2 = \sum_{i=1}^{141} \frac{(C_i - 62/141)^2}{62/141}$$

We compare the observed value of this statistic with the critical value from a chi-square table, with $141 - 1 = 140$ degrees of freedom. The data yield an observed statistic of 183.61; this exceeds the critical value of 168.6, and so the null hypothesis is rejected. The p-value associated with the observed test statistic is .014. It should be noted that even though the expectations are very small, the distributional assumption is accurate. This was verified by randomly placing 62 cases in the 141 regions; the 95th percentile of the simulated chi-square distribution was 170.0—very close to the tabled value of 168.6. The proportion of simulations exceeding the tabled critical value of 168.6 was 0.059, implying that use of the tables is just slightly liberal. Similarly, the simulations revealed that the probability of a statistic higher than the observed value of 183.61 was .009; this is just slightly lower than the value of .014 found from the tables, and also suggests that use of the tables is slightly liberal. It is remarkable how close the tabled values are to the simulated values, and this gives support to the use of tables in instances such as this where the expectations are small.

It is of interest to know which areas of the study region contribute most to the significant global statistic. Accounting for the multiplicity of testing, the critical value of a z-statistic used in 141 tests is equal to $\Phi^{-1}(1 - .05/141) = 3.39$, where Φ is used to indicate the cumulative distribution function of a standard normal random variable. The region around each control may be tested using the local statistic

$$z_i = \frac{c_i - 0.5 - 62/141}{\sqrt{62/141}}$$

This includes a correction for continuity. Results indicate that the following regions are significant, or close to significant. Two controls have three cases that are closer to the control than to any other control (compared with an expectation of 62/141 cases, yielding a z-value of 3.11). These two controls are at the following locations on Figure 3.7:

(5172, 4674)—Northern portion of the study area
(5116, 4308)—Middle of cluster of points

One control, located at (5150, 4303), has four cases that are nearer to it than to any other control ($z = 4.62$). This location is in the portion of the study area containing a dense cluster of points. It should be noted that caution should be used in interpreting the z-scores because they can be misleading when expectations are as small as they are.

When cases were used to generate polygons, the expected number of controls in each polygon is 141/62. The chi-squared value in this case was 119.75, which is significant with $p < .001$, using $62 - 1 = 61$ degrees of freedom. A case with coordinates (5030, 4299) contributes most strongly to the global statistic; it has 10 controls that are closer to it than to any other case. This might be considered a potential "cool spot."

3.11.2   Spatial Scale

Until this point, we have implicitly assumed a spatial scale of interest by defining polygons around each control, or around each case. A natural question to ask is whether larger spatial scales may be examined by generalizing the current approach to a more explicitly spatial statistic that makes use of the number of cases and controls in the neighborhood of each polygon. These extensions are not pursued here but are possible using the approaches adopted by Tango (1995) and Rogerson (1999). Lost in such generalization, however, is the simplicity that originally motivated this test.

3.12   A Modified Cuzick–Edwards Test

Rather than using the k-nearest neighbors to define the observed statistic as the number of case–case pairs, we now define the observed statistic as the number of case–case pairs that reside within a distance h of one another. This has a more natural affinity with the concept of defining a spatial scale for the potential cluster because for any case the k-nearest neighbors may be either close to, or very far from, the case. The remainder of the proposed approach is identical to that used by Cuzick and Edwards—one computes a z-score by subtracting the expectation of the number of case–case pairs from the test statistic and then dividing the result by the standard deviation.

A more formal approach to describing this in detail is by means of K-functions. From the definition of K-functions, the observed number of case pairs that lie within a distance h of one another is equal to

$$CC_{obs} = N_0^2 \hat{K}_{00}(h)/R \qquad (3.48)$$

where R is the size of the study area, N_0 is the number of cases, and $\hat{K}_{00}$ is the estimated K-function using the case locations. In turn,

$$\hat{K}_{00}(h) = \frac{R}{N_0^2} \sum_i \sum_{j \neq i} I_{ij}(h) \tag{3.49}$$

where $I_{ij}(h)$ is equal to one if points i and j are separated by a distance less than h, and is equal to zero otherwise.

We next compare this observed count with the expected count of pairs of cases within a distance h, under the null hypothesis of random labeling. This quantity is equal to

$$CC_{expected} = \frac{N_0 - 1}{N_0 + N_1 - 1} \left(\frac{N_0^2 K_{00}}{R} + \frac{N_0 N_1 K_{01}}{R} \right) \tag{3.50}$$

where K_{01} is the cross K-function, given by

$$\hat{K}_{01}(h) = \frac{R}{N_0 N_1} \sum_{i=1}^{N_0} \sum_{j=1}^{N_1} I_{ij}(h) \tag{3.51}$$

where $I_{ij}(h)$ is now equal to one if control j and case i are separated by a distance less than h, and is equal to zero otherwise.

To see how the expectation in Equation 3.50 is derived, recognize first that the term in parentheses is simply the total number of case and control locations within a distance h of a given case, summed over all cases. The first term in parentheses is a count of the number of case–case pairs within a distance h of one another, and the second term in parentheses is a count of the number of controls within a distance h of cases. Under random relabeling, a proportion of this total is expected to consist of case–case pairs. Because for each case there are $N_0 - 1$ other cases, and $N_0 + N_1 - 1$ other case and control locations, the desired proportion is $(N_0 - 1)/(N_0 + N_1 - 1)$.

The calculation of $CC_{expected}$ is given in terms of K-functions in (Equation 3.50), and may be described simply and directly as follows: first determine the number of case and control locations within a distance h of each case, and then sum over all cases. This result is then multiplied by $(N_0 - 1)/(N_0 + N_1 - 1)$ to obtain $CC_{expected}$.

The difference between observed and expected counts may be written as

$$CC_{obs} - CC_{exp} = \frac{N_0^2 N_1}{R(N_0 + N_1 - 1)} (K_{00} - K_{01}) + \frac{N_0 N_1 K_{01}}{R(N_0 + N_1 - 1)} \tag{3.52}$$

The first term on the right-hand side is dominant, and this suggests that it is reasonable to use the difference between the *K*-function for cases and the cross *K*-function as a basis for a test of clustering—a suggestion made by Kulldorff (1998) in noting the deficiencies associated with using the difference between *K*-functions for cases and controls ($K_{00} - K_{11}$) to test for spatial clustering. The idea of using the difference between *K*-functions was suggested by Diggle and Chetwynd (1991), and its limitations have also been noted by Tango (1999).

To use the quantity in Equation 3.52 for a statistical test of the null hypothesis of random labeling, it remains to determine its variance. As a first approximation, when the spatial scale *h* is small, the variance is approximately equal to the expectation. This was confirmed via simulations. Therefore, a *z*-statistic may be used as an approximation:

$$z = \frac{CC_{obs} - CC_{exp}}{\sqrt{CC_{exp}}} \tag{3.53}$$

For larger values of *h*, a more accurate test could be carried out by Monte Carlo simulation of the null hypothesis of no spatial clustering, with random relabeling of cases and controls. Such simulation would yield a more accurate distribution of the test statistic, CC_{obs}. An alternative would be to attempt to improve on the expression for the variance by deriving the covariance between the number of cases close to a specific case, and the number of cases close to another, nearby, case.

3.12.1 Example: Leukemia and Lymphoma Case–Control Data in England

Table 3.12 summarizes the results for different values of *h*, using the leukemia and lymphoma case–control data in Humberside, England; the data are briefly described in Chapter 1, subsection 1.7.2.5, and consist of 62 cases and 141 controls. Values of the test statistic for higher values of *h* were all insignificant.

TABLE 3.12

Results of Modified Cuzick-Edwards Test: Humberside Data

h	CC_{obs}	CC_{exp}	*z*	*p*-value
3	26	15.7	2.60	.005
4	32	24.2	1.60	.055
5	50	36.8	2.17	.015
6	60	45.0	2.24	.013
7	68	55.3	1.71	.043
8	76	67.3	1.06	.146
9	90	85.5	0.49	.317

TABLE 3.13

Local Sources of Contributions to Significant
Modified Cuzick-Edwards Statistic

$h = 3$

Case locations with 3 locations within h:

Case	x	y
58	5176	4669
59	5177	4667
60	5177	4667

$h = 5$

Case locations with 4 locations within h:

Case	x	y
9	5026	4300
10	5026	4301

$h = 6$

Case locations with 4 locations within h:

Case	x	y
9	5026	4300
10	5026	4301
58	5176	4669

For prespecified h values of 3, 5, 6, and 7, the null hypothesis is rejected; there are significantly more case pairs within these distances than would be expected. Locations contributing to the clustering at selected scales are in Table 3.13; these coordinates refer to those used in Figure 3.7.

In practice, h will not be known, and it will often be of interest to adjust for the multiple testing that occurs over a range of h values. One could simply apply a Bonferroni adjustment (Chapter 5, Section 5.8) based on the number of h values examined, but this would be conservative, owing to the correlation among tests with similar values of h. Ord (1990), in his discussion of the paper by Cuzick and Edwards, suggests looking at the *differences* in the results for successive h values. Because these differences are essentially independent, a Bonferroni adjustment could then be used.

4

Local Statistics

4.1 Introduction

As we have seen in Chapter 3, there are various global statistics that may
be employed to test the null hypothesis that a map pattern does not deviate significantly from expectations. These tests do not provide information
on the question of whether incidence may be raised significantly around
particular locations. In providing a single test statistic, the global tests do
not reveal information about the size, location, and significance of local-
ized geographic clusters.

When global statistics prove significant, it is of interest to know
which regions are responsible for the significance. Even when global
statistics are not significant, it is still possible that there are significant
local "pockets" of raised incidence; they just may not be substantial
enough in terms of raised risk and size to ensure that the global statistic is significant.

In this chapter, we review several *local* or *focused* statistics. Some
of these statistics were designed to test the null hypothesis of whether
observed counts are raised in the vicinity of a particular location, relative to the expected counts. For example, there is often interest in determining whether the number of observed health events in the vicinity of
putative sources is greater than what could be expected by chance alone.
Other local statistics were developed to test the null hypothesis that values observed for a variable are spatially independent in the vicinity of a
location of interest.

A desirable property of local statistics is that the sum of the local statistics, across all regions, be equal to a multiple of the global statistic. Thus,
the global statistic may be decomposed into a set of regional, local statistics. A good, early discussion of this property, as well as other features of
local statistics, may be found in Anselin (1995).

Score statistics, reviewed later in this chapter, are locally most powerful tests against specified alternative hypotheses. Lawson (1993) provides
a good overview of such tests and derives score statistics for a number of
alternatives (e.g., directional alternatives where incidence may be raised
in some geographic directions but not others).

4.2   Local Moran Statistic

The local Moran statistic is used to determine whether local spatial auto-correlation exists around a specified subregion i ($i = 1, ..., m$), and it is defined as

$$I_i = \frac{m(y_i - \bar{y})}{\Sigma_j (y_j - \bar{y})^2} \sum_j w_{ij}(y_j - \bar{y}) \tag{4.1}$$

where the notation is the same as in Chapter 3, Equation 3.10. The sum of the local Moran values obtained for all subregions in a study region is equal to the global Moran value multiplied by the sum of w_{ij}; that is, $I = \Sigma_i \Sigma_j w_{ij} \Sigma_i I_i$. Anselin (1995) gives the expected value and variance of I_i for the randomization hypothesis as follows:

$$E[I_i] = -\frac{w_i}{m-1} \tag{4.2}$$

$$V[I_i] = w_{i(2)}(m - b_2)/(m-1)$$
$$+ 2w_{i(kh)}(2b_2 - m)/(m-1)(m-2) - w_i^2/(m-1)^2 \tag{4.3}$$

where $w_i = \Sigma_j w_{ij}$, $w_{i(2)} = \Sigma_{j \neq i} w_{ij}^2$, $2w_{i(kh)} = \Sigma_{k \neq i} \Sigma_{h \neq i} w_{ik} w_{ih}$, and $b_2 = m\Sigma_i (y_i - \bar{y})^4 / \{\Sigma_i(y_i - \bar{y})^2\}^2$.

He also assesses the adequacy of the assumption that the test statistic has a normal distribution under the null hypothesis of no spatial auto-correlation. Through simulation study with datasets with $m = 42$ and $m = 81$, Anselin concludes that the normal distribution may be inappropriate to approximate the distribution of I_i, and the use of the normal approximation would lead to an overrejection of the null hypothesis, although this might be of lesser concern with larger sample sizes or a higher number of neighbors, or both. In addition, Anselin points out that the expected value and the variance given by Equations 4.2 and 4.3, respectively, are not appropriate when global spatial association exists because they are derived based on the assumption that each value is equally likely at any region. He thus suggests that a simulation process with a conditional randomization approach (in which the observed value y_i at region i is fixed, and the remaining $m -1$ values are randomly permuted over regions other than i) be taken to obtain pseudosignificance levels for I_i, regardless of the presence or absence of global spatial association.

4.2.1 Illustration

Let us consider the six-region system used in Chapter 3 (Figure 3.2), and assume that we are interested in local spatial autocorrelation around region 6. Assuming that a nonstandardized binary connectivity matrix is used to determine w_{ij}, $\sum_j w_{6j} = 2$; in addition, $m = 6$, $\bar{y} = \sum_i y_i/6 = 20$, and $\sum_i (y_i - \bar{y})^2 = 370$. For region 6, $y_6 - \bar{y} = 30 - 20 = 10$ and $\sum_j w_{6j}(y_j - \bar{y}) = (25 - 20) + (28 - 20) = 13$. So, the local Moran statistic for region 6 is calculated as

$$I_6 = \frac{m(y_6 - \bar{y})}{\sum_j (y_j - \bar{y})^2} \sum_j w_{6j}(y_j - \bar{y}) = \frac{6 \cdot 10}{370} \cdot 13 = 2.108$$

Using Equations 4.2 and 4.3, the expected value and the variance are $E[I_6] = -0.4$ and $V[I_6] = 1.04$, respectively. The z-score associated with the observed I_6 value is $z = \frac{2.108 - (-0.4)}{\sqrt{1.04}} = 2.459$, implying significant local positive spatial autocorrelation around region 6. Note, however, that the normal approximation of the distribution of I_6 is unlikely to be appropriate for this small data set, and the previous result is purely for the purpose of illustration.

4.2.2 Example: Low Birthweight Cases in California

The low birthweight (LBW) data for California in 2000 described in Chapter 1, subsection 1.7.2.6, is aggregated at the county level and then converted into z-scores under the assumption that LBW cases are distributed among all births following a Poisson distribution. That is, for each county i, the expected number of LBW cases E_i is obtained based on the total number of births observed in the county and an overall LBW rate that is constant over the study region; then, the observed number of cases O_i is converted into a z-score as $z_i = \frac{O_i - E_i}{\sqrt{E_i}}$. Figure 4.1a shows the distribution of z-scores. A general characteristic that can be seen in this map is that relatively few counties have positive z-scores (i.e., more LBW cases than expected), and they are somewhat concentrated around Los Angeles and San Francisco counties.

Local Moran's I statistic is then applied to this z-score distribution using a row-standardized rook's connectivity matrix. Figure 4.1b presents z-scores associated with the resulting local Moran's I values. Statistically significant positive spatial autocorrelation is found around Contra Costa, Alameda (both next to San Francisco), Kern (to the north of Los Angeles), and San Bernardino (to the east of Los Angeles). Because the four counties all have positive z-scores for the observed number of cases as shown in Figure 4.1a, the detected local spatial autocorrelation can be seen reflecting the situation in which a higher value is surrounded by higher values in the

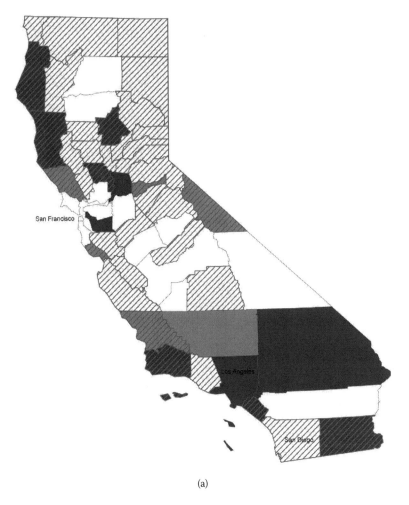

(a)

FIGURE 4.1
Low birthweight cases in California, 2000: (a) z-score of the observed number of LBW cases in relation to the total births (b) z-score associated with local Moran's *I*.

neighborhood, implying concentration of higher LBW risk around the four counties. Statistically significant negative spatial autocorrelation is detected only around Orange County, which has a negative z-score for the observed number of cases and is surrounded by counties with positive z-scores.

It should also be mentioned that no global spatial autocorrelation is detected in this dataset by global Moran's *I* analysis (*I* = 0.026 and z = 0.498). The detected local spatial autocorrelations can therefore be considered as local pockets of raised LBW risk. (Note that the example shown in Chapter 3, subsection 3.4.2, also used the California LBW data, but the Poisson-based standardization was not applied there.)

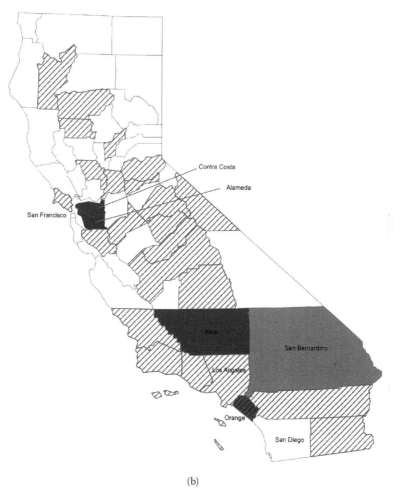

(b)

FIGURE 4.1
(Continued).

4.3 Score Statistic

The score statistic for testing the null hypothesis of no raised incidence around a prespecified site i is

$$U_i = n \left\{ \sum_{j=1}^{m} w_{ij}(r_j - p_j) \right\} \tag{4.4}$$

where r_j and p_j are the observed and expected proportions of all cases falling in region j ($j = 1, ..., m$), respectively; a_j is a weight representing exposure to the prespecified site by residents of region j; and n is the total number of cases found in the study region. The expectation of the statistic is zero, and the variance is

$$V[U_i] = n \left\{ \sum_{j=1}^{m} w_{ij}^2 p_j - \left(\sum_{j=1}^{m} w_{ij} p_j \right)^2 \right\} \qquad (4.5)$$

This can be used to form a z-statistic, which may then be compared with an appropriate critical value by assuming that the z-score is asymptotically normal under the null hypothesis of no raised incidence.

Waller et al. (1992) have shown that such a score test is uniformly most powerful against the focused clustering alternative:

$$H_1 : E[N_j] = np_j(1 + w_{0j}\varepsilon) \qquad (4.6)$$

where w_{0j} is a function of the distance from the focus o to region j, and ε is positive. $E[N_j]$ is the number of cases expected in region j. This alternative implies that the expected number of cases in region j is greater than that expected due to background population effects by an amount that depends on the distance from the focus. Lawson (1993) pays particular attention to alternative functions of distance in this context.

4.3.1   Illustration

Suppose we have the configuration in Figure 4.2, and wish to test the null hypothesis of no raised incidence in the vicinity of region 1. Suppose also

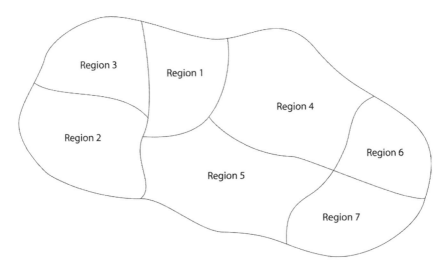

FIGURE 4.2
Hypothetical configuration of seven-region system.

TABLE 4.1

Observed and Expected Proportions of Cases in Each Region

Region	Observed Proportion	Expected Proportion	Weight
1	0.2	0.15	1
2	0.2	0.18	0.3
3	0.15	0.13	0.5
4	0.05	0.08	0.2
5	0.1	0.13	0.2
6	0.1	0.13	0.1
7	0.2	0.2	0.1

that we have a total of $n = 100$ cases and that the observed and expected proportions of these cases falling into the various subregions are as shown in Table 4.1. The score statistic for these data is

$$U = n \left\{ \sum_{j=1}^{m} w_{ij}(r_j - p_j) \right\} = 5.1$$

and the variance of U is found from

$$V[U] = n \left\{ \sum_{j=1}^{m} w_{ij}^2 p_j - \left(\sum_{j=1}^{m} w_{ij} p_j \right)^2 \right\} = 100(.2104 - .1184) = 9.2$$

Therefore, $z = 5.1/\sqrt{9.2} = 1.68$, and this is significant using a one-tailed test and a Type I error probability of .05.

4.4 Tango's C_F Statistic

Tango (1995) uses a modified and generalized score statistic to test for clusters around prespecified foci. His statistic is

$$C_F = \mathbf{c}'\mathbf{W}(\mathbf{r} - \mathbf{p}) \tag{4.7}$$

where $\mathbf{c}$ is an $m \times 1$ vector containing elements $c_i = 1$ if i is one of the prespecified foci, and 0 otherwise; $\mathbf{r}$ and $\mathbf{p}$ are $m \times 1$ vectors with elements containing the observed and expected proportions of cases in each region; and $\mathbf{W}$ is a matrix containing elements w_{ij}, which measure the closeness (or connectivity) between regions i and j. The variance of C_F is

$$V[C_F] = \mathbf{c}'\mathbf{W}\mathbf{V}_p\mathbf{W}\mathbf{c} \tag{4.8}$$

and under the null hypothesis of no pattern, the quantity $C_F^2/V[C_F]$ has a chi-square distribution with one degree of freedom. This statistic has the advantage of allowing more than one focal point to be specified simultaneously, and it also has been found to be quite powerful in rejecting false null hypotheses, especially when the number of prespecified foci is small.

4.4.1   Illustration

Again, consider the six-region system and use region 6 as a single prespecified focus, that is, $c = (0 \quad 0 \quad 0 \quad 0 \quad 0 \quad 1)'$. Further, assume that the proportion of the expected number of cases in the six region is given by $p = (0.1 \quad 0.1 \quad 0.1 \quad 0.3 \quad 0.3 \quad 0.1)'$.

From Chapter 3, Figure 3.2, the proportion of the observed cases in each region is $r = (0.117 \quad 0.100 \quad 0.092 \quad 0.208 \quad 0.233 \quad 0.250)'$. In addition, the weight matrix W is $W = \begin{pmatrix} 1 & 1 & 1 & 1 & 0 & 0 \\ 1 & 1 & 0 & 1 & 1 & 0 \\ 1 & 0 & 1 & 1 & 0 & 0 \\ 1 & 1 & 1 & 1 & 1 & 1 \\ 0 & 1 & 0 & 1 & 1 & 1 \\ 0 & 0 & 0 & 1 & 1 & 1 \end{pmatrix}$. The value of C_F is then obtained as $C_F = c'W(r - p) = -0.00833$.

Using Chapter 3, Equation 3.25,

$$V_p = \Delta p - pp' = \begin{pmatrix} 0.1 & 0 & 0 & 0 & 0 & 0 \\ 0 & 0.1 & 0 & 0 & 0 & 0 \\ 0 & 0 & 0.1 & 0 & 0 & 0 \\ 0 & 0 & 0 & 0.3 & 0 & 0 \\ 0 & 0 & 0 & 0 & 0.3 & 0 \\ 0 & 0 & 0 & 0 & 0 & 0.1 \end{pmatrix} - \begin{pmatrix} 0.1 \\ 0.1 \\ 0.1 \\ 0.3 \\ 0.3 \\ 0.1 \end{pmatrix}\begin{pmatrix} 0.1 \\ 0.1 \\ 0.1 \\ 0.3 \\ 0.3 \\ 0.1 \end{pmatrix}'$$

$$= \begin{pmatrix} 0.09 & -0.01 & -0.01 & -0.03 & -0.03 & -0.01 \\ -0.01 & 0.09 & -0.01 & -0.03 & -0.03 & -0.01 \\ -0.01 & -0.01 & 0.09 & -0.03 & -0.03 & -0.01 \\ -0.03 & -0.03 & -0.03 & 0.21 & -0.09 & -0.03 \\ -0.03 & -0.03 & -0.03 & -0.09 & 0.21 & -0.03 \\ -0.01 & -0.01 & -0.01 & -0.03 & -0.03 & 0.09 \end{pmatrix}$$

Then the variance of C_F is given by $V[C_F] = c'AV_pAc = 0.21$.

Finally, $C_F^2/V[C_F] = (-0.00833)^2/0.21 = 0.00033$, for which the p-value is .985 with one degree of freedom. This result implies that the observed concentration of cases in region 6 and its vicinity is not significant at all when the unequal distribution of the expected numbers of cases (which often reflects an unequal distribution of the risk population) is taken into account.

4.5   Getis' G_i Statistic

Ord and Getis (1995) have used the statistic

$$G_i = \frac{\sum_j w_{ij}(d)x_j - W_i\bar{x}}{s(i)\left\{\left[(m-1)S_{1i} - W_i^2\right]/(m-2)\right\}^{1/2}} \quad j \neq i, \text{ and}$$

$$G_i^* = \frac{\sum_j w_{ij}(d)x_j - W_i^*\bar{x}}{s\left\{\left[mS_{1i}^* - W_i^{*2}\right]/(m-1)\right\}^{1/2}} \quad \text{all } j$$

(4.9)

where

$$W_i = \sum_{j \neq i} w_{ij}(d)$$

$$W_i^* = W_i + w_{ii} = \sum_j w_{ij}(d)$$

$$S_{1i} = \sum_{j \neq i} \{w_{ij}(d)\}^2$$

$$S_{1i}^* = \sum_j \{w_{ij}(d)\}^2$$

(4.10)

The element $w_{ij}(d)$ is a spatial weight reflecting the proximity between regions i and j, and is generally set to one if region j is within a distance of d from region i, and 0 otherwise. Also, $\bar{x}$ and s are the sample mean and standard deviation of the observed set of x_i ($i = 1, ..., m$), respectively, and $s^2(i) = \sum_j x_j^2/(m-1) - \{\sum_j x_j/(m-1)\}^2$.

The only difference between the two variants of the local G statistic is whether or not the target region i is included in the computation of the statistic. Although Ord and Getis do not provide any specific explanation about in what situations one is more suitable than the other, G_i^* intuitively appears to be a more natural measure of local clustering tendency than G_i. The latter may detect as a cluster a region with a low value but surrounded by high-valued regions, which is often unlikely to be the kind of "cluster" one wants to detect. However, G_i is useful when a target region i has a substantially high value in comparison with its surrounding regions because the high value of the target region, if included, may make almost everything significant.

Ord and Getis note that, when the underlying variable has a normal distribution, so does the test statistic. Furthermore, when the underlying distribution is not normal, the distribution is asymptotically normal as the distance d becomes large. Because the statistic is written in standardized form, it can be taken as a standard normal random variable, with mean 0 and variance 1. However, Anselin (1995) notes that the assumption of the normally distributed test statistic may not hold under the presence of global spatial association, as is the case with the local Moran statistic.

4.5.1   Illustration

For the six-region system used in the previous section, let us assume that d is chosen so that only regions 4 and 5 are within d of region 6. For region 6, $W_6^* = 2$ and $S_{16}^* = 2$, the mean and the standard deviation of the observed

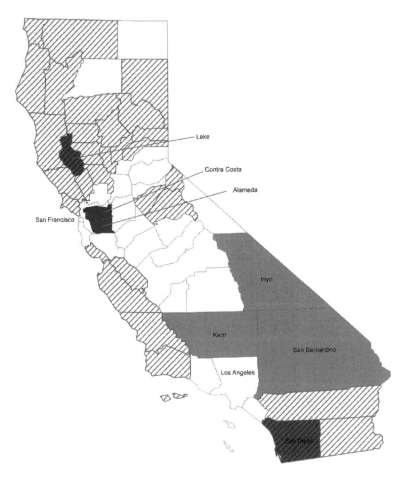

FIGURE 4.3
z-scores associated with the local G^* statistic applied to the low birthweight data in California, 2000.

6 regional values are $\bar{x} = 20$ and $s = 7.853$, respectively. Then, the local G_i^* statistic is given by

$$G_6^* = \frac{\sum_j w_{6j}(d)x_j - W_6^*\bar{x}}{s\left\{\left[mS_{16}^* - (W_6^*)^2\right]/(m-1)\right\}^{1/2}} = \frac{(25+28) - 2 \cdot 20}{7.853[(6 \cdot 2 - 2^2)/(6-1)]^{1/2}} = 1.309$$

In this example, the regional values do not have a normal distribution, so that some simulation process would be needed to evaluate the significance of the observed local statistic.

4.5.2 Example: Low Birthweight Cases in California

The local G_i^* is applied to the California LBW data used in subsection 4.2.2, and the distribution of z-scores associated with the resulting G_i^* values is presented in Figure 4.3. The four counties around which positive spatial autocorrelation was found in subsection 4.2.2 (i.e., Contra Costa, Alameda, Kern, and San Bernardino) and Inyo County have significantly positive z-scores here, indicating concentration of higher LBW risk around them. On the other hand, Lake (to the north of San Francisco) and San Diego (to the south of Los Angeles) counties have significant negative local G_i^* values, indicating concentration of lower LBW risk in these areas.

4.6 Stone's Test

For Stone's (1988) test, the regions are first ordered in terms of distance away from the hypothesized source (i.e., region 1), and a table of the cumulative observed and expected number of cases is formed as shown in Table 4.2, which is based on the data set shown in Table 4.1 and Figure 4.2. Stone's

TABLE 4.2

Results for Stone's Test

Region	Cumulative Observed Cases	Cumulative Expected Cases	Observed/Expected
1	20	15	1.33
3	35	28	1.25
2	55	46	1.20
4	60	54	1.11
5	70	67	1.04
6	80	80	1.0
7	100	100	1.0

Note: Regions are ordered in terms of distance from region 1.

test is based on the likelihood of the maximum ratio of the cumulative observed number of cases to the cumulative expected number of cases. This quantity is then compared with the distribution of that ratio, which in turn is found by simulating the null hypothesis.

4.6.1   Illustration

The maximum observed/expected ratio in Table 4.2 is 1.33, which is found in the first row. To assess the likelihood of this maximum observed/ expected ratio under the null hypothesis of no raised incidence in the neighborhood of region 1, simulation was conducted that distributed 100 cases to regions according to the expected proportions. Approximately 95% of the time (among 10,000 repetitions), the maximum ratio was equal to 1.40 or less. This implies that we should accept the null hypothesis at the 0.05 level. Among the 10,000 simulations, 1,081 had observed maximum ratios greater than 1.33, implying a *p*-value of .1081.

4.7   Modeling around Point Sources with Case–Control Data

In addition to nonparametric tests (such as Stone's, reviewed in the previous section), parametric approaches making use of point process models (Diggle 1990; Lawson 1993; Diggle and Rowlingson 1994) have been suggested to assess raised incidence around a point source.

   Diggle and Rowlingson (1994) have suggested a likelihood approach for testing retrospectively the null hypothesis of no raised incidence around a prespecified location, when data on the locations of cases and controls are available. Using their notation, assume that data are available on the locations on n cases and l controls. The intensity of disease $\lambda(\mathbf{x})$ at a location $\mathbf{x}$ is modeled as

$$\lambda(\mathbf{x}) = \rho\lambda_0(\mathbf{x})f(\mathbf{x} - \mathbf{x}_0;\theta) \tag{4.11}$$

where $\lambda_0(\mathbf{x})$ represents background intensity due to the population at risk, and ρ is a scaling parameter reflecting the ratio of the number of cases to that of controls. Furthermore, risk at location $\mathbf{x}$ is presumed to vary with location according to the function $f(\mathbf{x} - \mathbf{x}_0; \theta)$, where $\mathbf{x}_0$ is the prespecified location and θ is a set of parameters. They suggest the following function:

$$f(\mathbf{x} - \mathbf{x}_0;\theta) = 1 + \theta_1 e^{-\theta_2 d^2} \tag{4.12}$$

where d^2 is used to indicate the squared distance between locations $\mathbf{x}$ and $\mathbf{x}_0$. The parameter θ_1 estimates the excess risk at the source, and the parameter θ_2 represents exponential decline in risk as one travels away from the source. Other specifications for f that may be deemed more appropriate could also be adopted.

Conditional on the locations, the probability that an event at $\mathbf{x}$ is a case is

$$p(\mathbf{x}) = \frac{\rho f(\mathbf{x} - \mathbf{x}_0; \theta)}{1 + \rho f(\mathbf{x} - \mathbf{x}_0; \theta)} \tag{4.13}$$

The likelihood of the observed sample of cases and controls is given by

$$L(\rho, \theta) = n \ln \rho + \sum_{i=1}^{n} \ln f(\mathbf{x}_i - \mathbf{x}_0; \theta) - \sum_{i=1}^{n+l} \ln\{1 + \rho f(\mathbf{x}_i - \mathbf{x}_0; \theta)\} \tag{4.14}$$

When the null hypothesis of no raised incidence around the prespecified point is true and when Equation 4.12 is used to model the relationship between location and risk, with $\theta_1 = \theta_2 = 0$, and the likelihood equation, Equation 4.14, reduces to

$$L_0(\rho) = n \ln \rho - (n + l) \ln(1 + \rho) \tag{4.15}$$

Under the null hypothesis, Equation 4.15 is maximized at $\hat{\rho} = n/l$, and thus

$$L_0(\hat{\rho}) = n \ln(n/l) - (n + l) \ln\left(\frac{n + l}{l}\right) \tag{4.16}$$

A formal test of the null hypothesis is carried out by finding the parameter estimates that maximize Equation 4.14, and then comparing the quantity $D = 2\{L(\hat{\rho}, \hat{\theta}) - L_0(\hat{\rho})\}$ with the critical value of a χ^2 distribution having degrees of freedom equal to the number of parameters in θ; for instance, the degrees of freedom will be 2 if Equation 4.12 is used.

4.8 Cumulative and Maximum Chi-Square Tests as Focused Tests

Suppose that cases and controls are arranged within m zones, and that the zones are ordered according to distance away from a prespecified location of interest. Cumulative and maximum chi-square tests may be used to test the null hypothesis that cases and controls are randomly labeled.

The cumulative chi-square statistic is defined as

$$T = \sum_{k=1}^{m-1} \chi_k^2 \tag{4.17}$$

where χ_k^2 is the chi-square statistic for the 2×2 table consisting of columns representing the number of cases and controls, respectively. The two rows of the table represent the total of cases and controls found from aggregating the regions away from the source into two categories—from zone 1 to zone k, and $k + 1$ to zone m. A high value of T implies that there are large discrepancies between the distributions of cases and controls over a range of distances, resulting in rejection of the null hypothesis.

Hirotsu (1993) calculates the p-value associated with T from the quantities

$$d = 1 + \frac{2}{m-1}\left(\frac{\lambda_1}{\lambda_2} + \frac{\lambda_1 + \lambda_2}{\lambda_3} + \cdots + \frac{\lambda_1 + \cdots + \lambda_{m-2}}{\lambda_{m-1}} \right)$$

$$v = \frac{m-1}{d} \tag{4.18}$$

$$\lambda_k = \frac{N_1 + \cdots + N_k}{N_{k+1} + \cdots + N_m}$$

where N_i is the total number of observations (cases plus controls) at ordered zone i. Under the null hypothesis of random labeling, T/d has a distribution that is approximated by a chi-square distribution with v degrees of freedom.

Although Lagazio et al. (1996) note the utility of the cumulative chi-squared tests in testing the null hypothesis of no raised incidence around a prespecified site, there are two additional useful points that can be made regarding its implementation. The first is that if information on case–control locations is available, there is no need to aggregate cases and controls into zones. If there are C cases and controls around the prespecified point, there can be C zones, with exactly one case or control in each zone. This is because it is acceptable to use expectations even less than one when using the chi-squared test (Koehler and Larntz 1980). (However, if the ratio of controls to cases is larger than about three or four, the expectations of the number of cases in each zone will be lower than that recommended.) A second point is that the distribution of the cumulative chi-squared statistic is particularly straightforward to specify where there is a large number of zones. Specifically, when the number of zones is large, the degrees of freedom approach 3.45. The asymptotic convergence to this value is relatively rapid; for $m = 30$ zones, the degrees of freedom (df) are equal to 3.22; for $m = 50$, $df = 3.31$; and for $m = 100$, $df = 3.38$. This may be verified using the following more precise approximation for degrees of freedom given by Hirotsu (1993):

$$df = \frac{(m-1)^2}{2m^2 \sum_{i=1}^{K} \frac{1}{i^2} - (3m^2 - 2m + 1)} \tag{4.19}$$

Thus, when the number of zones (m) is large, the statistic T/d has a chi-squared distribution with $v = 3.45$ degrees of freedom, where T is the cumulative chi-squared value and $d = (m - 1)/3.45$. Note that the statistic

T/d may be rewritten simply as $\frac{3.45T}{m-1}$. An alternative is to use the maximum chi-square statistic, defined as

$$S = \max_{1 \le k \le m-1} \chi_k^2 \qquad (4.20)$$

Conceptually, the data are again viewed as potentially coming from two (spatial) regimes—one including zones 1 through k, and the other including zones $k+1$ through m. A high value of S leads to rejection of the null hypothesis that the locations have been assigned case/control status at random (i.e., random labeling), in favor of the hypothesis that there are two spatial regimes with different case/control ratios.

Derivation of p-values for the maximum chi-square statistic S is discussed by Srivastava and Worsley (1986). Specifically, the probability that the test statistic is exceeded under the null hypothesis is approximately

$$\Pr(S > S_{obs}) \approx 1 - F_1(S_{obs}) + q_1 \sum_{r=1}^{m-2} t_r - q_2 \sum_{r=1}^{m-2} t_r^3 \qquad (4.21)$$

where

$$q_1 = 2\left(\frac{S_{obs}}{\pi}\right)^{0.5} f_1(S_{obs}) \qquad (4.22)$$

$$q_2 = q_1\left(\frac{S_{obs}-1}{12}\right) \qquad (4.23)$$

$$t_r = (1-\rho_r)^{1/2} \qquad (4.24)$$

$$\rho_r = \sqrt{\frac{\lambda_r}{\lambda_{r+1}}} \qquad (4.25)$$

and $f_1(S_{obs})$ and $F_1(S_{obs})$ are the probability density and cumulative density functions for a chi-square random variable with one degree of freedom, respectively. In addition, λ is defined as in Equation 4.18.

4.8.1 Illustration

Suppose $m = 5$, and there are 10, 13, 26, 20, and 15 fixed locations in zones 1 through 5, respectively. Each of the 84 locations is either a case or a control, and suppose that there are 46 cases and 38 controls in total. Ten thousand

Monte Carlo simulations of random labeling led to 95th percentiles of 10.22 and 5.818 for T and S, respectively.

For the cumulative chi-square statistic, we have $\lambda = \{\lambda_1, \ldots, \lambda_4\} = \{10/74, 23/61, 49/35, 69/15\}$. This leads to $d = 1.57$ and $v = 2.54$. The observed test statistic T of 10.22 is first be divided by 1.57; the result (7.05) is assessed using a chi-square distribution with 2.54 degrees of freedom. This reveals a p-value of .053, which is close to the "expected" value of 0.05 associated with the 95th percentile found by simulation.

For the maximum chi-square statistic, we first find

$$\rho_{12} = \sqrt{\frac{10/74}{23/61}} = .598; \quad \rho_{23} = \sqrt{\frac{23/61}{49/35}} = .519; \quad (4.26)$$

$$\rho_{34} = \sqrt{\frac{49/35}{69/31}} = .793$$

and $t_1 = .634$, $t_2 = .694$, and $t_3 = .455$. Then using Equation 4.21,

$$\Pr(S > 5.818) = 1 - .9842 + (2\sqrt{5.689/3.1416})^* \ .00902^* \ (.634 + .694 + .455)$$

$$- \{(2\sqrt{5.689/3.1416})^* \ .00902\}\{.634^3 + .694^3 + .455^3\}$$

$$= .0528$$

$$(4.27)$$

which is again near the expected value of 0.05.

4.8.2   Example: Leukemia and Lymphoma Case-Control Data in England

For the Humberside data explained in Chapter 1, subsection 1.7.2.5, $m = 203$. The critical value for the test statistic is found by noting first that $3.45T/202$ has a chi-square distribution with 3.45 degrees of freedom. The critical value for a chi-square distribution with this many degrees of freedom is 8.58. Approximately 10,000 simulations of the null hypothesis using random labeling were carried out. Comparing the simulated values of the quantity $3.45T/202$ with the critical value of 8.58 resulted in rejection of the null hypothesis in 4.5% of the simulations. This implies that the formula for the cumulative chi-square appears to be quite accurate and just slightly conservative.

Setting aside issues associated with multiple testing for the moment, searching all case and control observations as potential focal points leads to no rejections of the null hypothesis. The maximum value among the observed cumulative chi-squared results is 417.3; comparing the quantity $(3.45)417.3/202 = 7.13$ with a chi-square distribution with 3.45 degrees of freedom leads to $p = .09$ (and note that this is the most significant among the total 203 local statistics).

4.8.3 Discreteness of the Maximum Chi-Square Statistic

Inspection of the values of the maximum chi-square statistic that are simulated under the null hypothesis reveals that a small number of values dominate the upper tail of the distribution. In the present example, one run of 1000 simulations produced values of 5.689 12 times; on the ordered list of simulated statistics, these occupied positions 937–948. Similarly, 5.818 appeared 13 times; on the ordered list of statistics, these occupied positions 949–962. Simply inspecting the 950th value could therefore be a bit misleading; it can potentially be difficult to calculate exact significance levels because of the multiplicity of these simulated values. One alternative is to run such simulations repeatedly and note where on the list of ordered simulated statistics a change occurs from one value to another. In this case, the change from 5.689 to 5.818 always occurs near position 950, and on average occurred at position 947. Thus, we can take 5.689 as a critical value for a Type I error probability of $1 - 0.947 = .053$, as this value is exceeded under the null hypothesis approximately 5.3% of the time.

4.8.4 Relative Power of the Two Tests

To examine the relative power of the maximum and cumulative chi-square tests, alternative hypotheses were simulated for the data used in section 4.8.1 by randomly and sequentially assigning cases to unassigned case and control locations. This was done by

1. Multiplying the number of unassigned locations remaining in each of the first k zones by a factor $r > 1$. (The number of unassigned locations remaining in the last $m - k$ zones is multiplied by 1.)
2. The results of step 1 are divided by the sum of the results in step 1 for all zones to find zone-specific probabilities for the assignment of the next case.

Results of simulations for different values of k are shown in Table 4.3; the scaling factor r was set to 2. As expected, the likelihood of detecting a change is greater when the change occurs in the middle of a sequence (i.e.,

TABLE 4.3

Power of Cumulative and Maximum Chi-Square Tests

	Cumulative Chi-Square	Maximum Chi-Square
$k = 1$	0.185	0.141
$k = 2$	0.421	0.407
$k = 3$	0.457	0.449
$k = 4$	0.227	0.336

$k = 2$ or 3 in the present example), than when it occurs near the beginning or end of the sequence (in this case, $k = 1$ or 4).

In this example, the maximum chi-square test is best when the change comes late in the sequence (i.e., when there is a large cluster in the disease clustering case). For a change occurring in any of the first three zones, the cumulative chi-square test is better.

The maximum chi-square statistic has the advantage of suggesting that the change occurred after location k. However, examination of the simulated alternatives when $k = 1$ reveals that, when the maximum chi-square statistic is significant, the accompanying value of k is often incorrect. This is not true when $k = 4$; in that case, the maximum chi-square statistic often occurs for $k = 4$. It would be interesting to undertake a more complete comparison of these situations because the comparison here only begins to be suggestive of the nature of the test's differences.

4.9 The Local Quadrat Test and an Introduction to Multiple Testing via the *M*-Test

With the assumption that cell counts follow a Poisson distribution, a local statistic for a particular cell i may be constructed as follows. When cell counts follow a Poisson distribution, the probability that the observed count in an individual cell O_i is equaled or exceeded under the null hypothesis is

$$\sum_{x=O_i}^{\infty} \frac{e^{-E_i} E_i^x}{x!} \tag{4.28}$$

where E_i is the expected number of cases in cell i.

If the expectation is sufficiently large (greater than about five), an approximate test is

$$z_i = \frac{O_i - E_i}{\sqrt{E_i}} \tag{4.29}$$

and z_i can be taken to have a distribution that is approximately standard normal. This is an example of a local test in which the weights associated with all other regions are equal to zero. Any clustering is therefore expected at the scale of the cell or region, or within the region; this would not be a powerful way of detecting raised incidence that occurs over a scale of neighboring regions.

It will often be the case that we are interested in evaluating simultaneously the significance of many such local cases. In this event, the multiple testing of local statistics must be accounted for; otherwise, one may find that a local statistic displays "significance" when in fact it may have arisen by chance alone, as so many local statistics were tested. The next chapter focuses upon this scenario, where a map is scanned for local statistics that may point to the existence of significant local clusters. This is of course the third category of clustering in Besag and Newell's classification—tests for the detection of clustering. In the next subsection, we present an introduction to these tests for the case of local quadrat tests that are carried out on the scale of individual regions.

4.9.1 Fuchs and Kenett's *M* Test

Consider the case where a global chi-square test is carried out to compare observed and expected values. When the global test is significant, it is of interest to know which regions contributed to the significant value. Perhaps it was simply a region or two that might then be considered as potential outliers. Also, when the global test is insignificant, it is still possible that an examination of individual regions will uncover instances in which the observed value differs significantly from the expected value (perhaps considered local hot spots). Fuchs and Kenett (1980) propose the *M*-test to find outlying cells in the multinomial distribution. The test can be used in conjunction with the global chi-square test (and the related quadrat test). Their method is based upon a comparison of the most unusual cell (as measured by the highest among the individual z-scores) with an appropriate critical value that takes into account the multiple testing. More specifically, for a study region consisting of m grid cells, a z-score for an individual cell, i ($i = 1, \ldots, m$), is first defined below:

$$z_i = \frac{n_i - Np_i}{\sqrt{Np_i(1 - p_i)}} \tag{4.30}$$

where n_i is the number of observed cases in cell i, p_i is the expected proportion of cases falling in cell i, and N is the total number of cases observed in the study area. Note that Equation 4.30 is based on a binomial model for the regional counts; Equation 4.29, on the other hand, is based on a Poisson model but the same logic can be applied. Then a one-sided test is carried out based on $\max |z_i|; i = 1, \ldots, m$, using bounds for the critical value

$$\Phi^{-1}\left\{1 - \frac{m - [m^2 - 2\alpha m(m-1)]^{1/2}}{m(m-1)}\right\}, \quad \Phi^{-1}\left\{1 - \frac{\alpha}{m}\right\} \tag{4.31}$$

where α is the significance level, Φ is the cumulative distribution function for the standard normal distribution and $\Phi^{-1}(x)$ is the standard normal variate associated with an area of $1 - x$ in the right tail of the distribution. They show that the test is more powerful than the global chi-square test in the presence of a single outlier. The null hypothesis of no outlier is rejected when

TABLE 4.4

Expected Number of Points and Number of Points Needed
for Significance with the *M*-Test for Alternative Grid Sizes

	$\Phi^{-1}(1 - .05/n^2)$	Expected No. of Pts.	No. of Pts. Needed for Significance
2×2	2.24	20	30
3×3	2.54	8.9	16.5
4×4	2.73	5.0	11.1
5×5	2.88	3.2	8.3
6×6	2.99	2.22	6.7
7×7	3.08	1.63	5.6
8×8	3.16	1.25	4.8
9×9	3.23	0.99	4.2
10×10	3.29	0.80	3.7

the maximum z-score exceeds this critical value. This procedure implies that the m-test focuses only on the most extreme observation and evaluates it on the basis of the Bonferroni-adjusted significance level. Detailed discussion on the Bonferroni adjustment is given in Sections 5.2 and 5.3. Thus the *M*-test can be used in conjunction with the global test; when the global test is significant, the *M*-test will be useful in determining whether the global significance is achieved via contributions from many regions, or whether there is a small number of outliers. Similarly, when the global test is not significant because the *M*-test is more powerful in detecting outliers, the latter may uncover anomalous values that remain hidden in the global statistic.

To continue the illustration of Chapter 3, subsection 3.3.6, where there are 80 points in the study area, Table 4.4 gives, for each of our grids with various sizes, the expected number of points in each cell as well as the minimum number of points needed in a cell to reach significance. For example, cells in a 5×5 grid would contain, on average, 3.2 points, and if any cell contained 9 or more points, this would be taken as evidence of a significant outlier and a rejection of the hypothesis that the points were randomly distributed.

Table 4.5 shows the power of the *M*-test as well as the usual (global) χ^2 test to reject the null hypothesis for 10,000 simulations of the alternative hypothesis, where 80 points are placed in a study area, some at random with probability .9, and some randomly within a square at the middle of the study area bounded by x and y coordinates of {0.4, 0.6} with probability .1. Note that the *M*-test has power exceeding that of the usual χ^2 test when the grid size and location matches that of the cluster (i.e., the 5×5 grid). For other grid sizes (especially finer grid sizes) too, the *M*-test in general displays better power.

Because of the discreteness of the random variable, it is not possible to achieve a Type I error probability of precisely .05, and this makes comparisons somewhat difficult. In constructing Table 4.5, the last column of

TABLE 4.5

Comparison of Power for Global χ^2 and
M-Tests for Alternative Grid Sizes

Grid	χ^2	M
2×2	0.053	0.016
3×3	0.356	0.336
4×4	0.117	0.130
5×5	0.597	0.657
6×6	0.191	0.198
7×7	0.283	0.361
8×8	0.275	0.374
9×9	0.227	0.183
10×10	0.338	0.501

Table 4.4 has been rounded up, and hence the results reported in Table 4.5 for the M-test are conservative. For example, in the 9×9 grid, we require 4.2 points in a cell for significance; by requiring 5 points in a grid cell, we are actually using a Type I error probability of only .002, instead of the nominal value of .05.

4.9.2 Example 1: Sudden Infant Death Syndrome in North Carolina

For the sudden infant death syndrome (SIDS) data described in Chapter 1, subsection 1.7.2.3, and used in Chapter 3, subsection 3.3.2.2, we use $\alpha = 0.05$, and $m = 100$. This leads to respective lower and upper bounds of 3.283 and 3.291 for the maximum of the local statistics. There are three counties that exceed this (when z-scores are found by using Equation 4.30): county 4 ($z = 6.65$), county 78 ($z = 3.815$), and county 66 ($z = 3.623$). If these three counties had not been included in the original global analysis, the observed chi-square statistic would have been 154.2, with 96 degrees of freedom. It is equivalent to a standard normal deviate of 3.74 (equal to $\sqrt{2(154.2)} - \sqrt{2(96) - 1}$). This is still significant, though much lower than the original deviate of 7.21. We may conclude that the outliers have had a noticeable effect, though the deviation between observed and expected county frequencies cannot be attributable to these counties alone.

The null hypothesis of no outlier is rejected when the maximum z-score exceeds this critical value. This procedure implies that the M test focuses only on the most extreme observation and evaluates it on the basis of Bonferroni-adjusted significance level. Detailed discurssion on the Bonferroni adjustment is given in Sections 5.2 and 5.3.

4.9.3 Example 2: Lung Cancer in Cambridgeshire

For the data on lung cancer in Cambridgeshire used in Chapter 3, subsection 3.3.2.3, the maximum local value of $z = (O_i - 0.5 - E_i)/\sqrt{E_i}$ is 3.24

(where 0.5 represents a continuity correction); this corresponds to ward 73, where there were 5 observed cases and an expectation of 1.1 cases. The critical value is calculated by finding the z-score associated with an area of 0.05/157 in the upper tail of a standard normal distribution; this critical value is equal to 3.42, and hence, the extreme z-score is not significantly greater than would be expected under the null hypothesis.

5

Tests for the Detection of Clustering, Including Scan Statistics

5.1 Introduction

The application of local spatial statistics (Anselin 1995; Getis and Ord 1992; Ord and Getis 1995), as discussed in Chapter 4, is rapidly becoming more widespread in the field of spatial analysis. Often there is interest in searching for significant spatial associations with no or little prior information about their location. The third category of tests in Besag and Newell's classification, namely, tests for the detection of clustering, serve this purpose and essentially carry out multiple local tests of the null hypothesis of no raised incidence across the entire study region. These tests "scan" the study region to find subareas that constitute spatial clustering over and above what could be expected by chance alone. Often, such scanning is done not only across all locations within the study area, but also at different spatial scales.

An open problem in the use of local spatial statistics is how to properly account for the multiple local tests of significance that are carried out for many regions simultaneously. If a large number of local tests are carried out, one will almost certainly obtain some "significant" results by chance alone. For instance, if 1000 local tests are made, each using $\alpha = 0.05$, one would expect 50 of the 1000 test statistics to lie beyond their critical values merely by chance when their associated null hypotheses are actually true. Clearly, an adjustment to the usual method is needed if one is to avoid finding so many spurious "significant" results; this issue is generally referred to as the *multiple testing problem*. When the scan tests examine multiple spatial scales in addition to multiple locations in the study region, the multiplicity of the scales must also be adjusted for.

This chapter is organized as follows. In the next six sections (Sections 5.2–5.7), we review a variety of scan-type tests for the detection of clusters that were proposed in the previous two decades, with some examples. Then in Sections 5.8 and 5.9, the Bonferroni and Sidak adjustments for multiple testing are discussed; these constitute common and simple approaches to the multiple testing problem, but can be conservative in the sense that they do not reject the null hypothesis often enough. That is, when the null hypothesis is true, such tests reject it with a probability less than α. When the null hypothesis is not true, the more conservative the test, the lower the statistical power and the higher the probability of committing a Type II error by accepting a false null hypothesis. Section 5.10 discusses a kernel-based cluster detection method developed by Rogerson (2001a), along with a possible technique to avoid making a test too conservative by taking into account correlation between observations.

5.2   Openshaw et al.'s Geographical Analysis Machine (GAM)

Let us start our discussion of a variety of scan-type methods for the detection of clusters with Openshaw et al.'s (1987) Geographical Analysis Machine (GAM). With this exploratory method for detecting clusters, a grid of points is constructed over a study region. At each grid point, circles of various radii are constructed as shown in Figure 5.1. The number of incidents in each

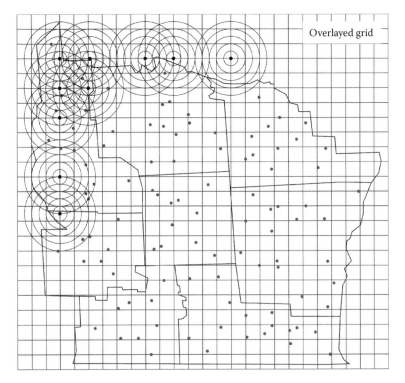

FIGURE 5.1
Openshaw's Geographical Analysis Machine. For each grid point, a set of circles with various radii are drawn and the number of points (i.e., incidents) falling in each circle is evaluated. To avoid too crowded a map, circles are not shown for all grid points.

circle is counted and compared with the number of incidents that would be expected if the pattern was random with respect to the underlying structure of population at risk. Although Openshaw originally suggested Monte Carlo testing at this stage, as Besag and Newell note, this is unnecessary, and a Poisson test could be used instead. If the actual number significantly exceeds the expected number, the circle is drawn on the map. The result is a map with a set of circles, where each circle has passed a test of significance.

Because many tests are carried out (that is, the number of grid points multiplied by the number of radii examined, which could be huge!), it is difficult to correct adequately for multiple tests. If one uses a conservative correction such as the Bonferroni adjustment, which essentially divides the overall significance level α by the number of tests conducted, it will be very difficult to find any clusters. Details of the Bonferroni adjustment will be provided later in this chapter. If, on the other hand, the degree of correction is not sufficient, clusters may be produced merely by chance. Openshaw uses a significance level of 0.002, but this is clearly arbitrary. The significance level used will dictate the number of circles plotted. In addition to the problem of multiple testing, a potential disadvantage of GAM is the enormous computational workload.

5.3 Besag and Newell's Test for the Detection of Clusters

Assume that the number of cases (or incidents) and the associated population at risk are available for each zone in the study area. Also assume that a critical number of cases, k, representing the size of clusters to be detected, is decided upon a priori. For a given case, i, neighboring zones are ordered in terms of increasing distance away from i, according to the locations of zone centroids. In Figure 5.2, numbers in parentheses represent the order with respect to the central (shaded in the figure) zone. The statistic T is the minimum number of nearest zones around case i that are needed to accumulate at least k cases. For example, for $k = 3$, $T = 3$ for the central zone in Figure 5.2. If T is small, that is indicative of a cluster around the zone containing case i.

Besag and Newell (1991) use their method for detecting clusters of rare diseases to find the likelihood that an even smaller number of zones could contain k cases, if the distribution of cases throughout the population was homogeneous. For an observed value t of T, the probability of $T \leq t$ can be approximated by equating it to the probability of k or more cases in the set of t zones, using Poisson probabilities. This in turn is equal to one minus the probability that the t zones contain k-1 or fewer cases.

$$\Pr(T \leq t) = 1 - \Pr(T > t) = 1 - \sum_{s=0}^{k-1} \exp(-u_t p)(u_t p)^s / s! \qquad (5.1)$$

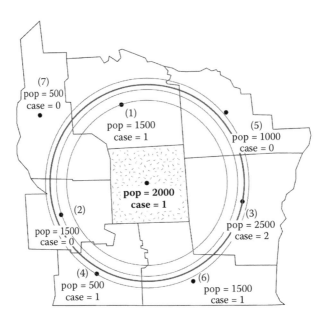

FIGURE 5.2
Besag and Newell's test for the detection of clusters. Dots in individual zones represent their centroids, and the circles are centered on the centroid of the central zone. The second largest circle contains three cases, excluding the one in the center zone, so that for $k = 3$, $T = 3$. Similar sets of circles should be examined for the rest of the zones. The population and the number of cases are hypothetical, not corresponding to any real data.

where u_t is the total population in the set of t zones and the central zone and p is the overall disease rate. A useful diagnostic of results suggested by Besag and Newell is a plot of all circles containing T zones that attained a given level of significance.

Although this method is computationally much less intensive than GAM, the results strongly depend on the arbitrary choice of k.

To this point, the description of this method pertains to clustering around a given region or case. Full implementation entails repeating this for each centroid. To fully account for multiple testing, one needs to realize that even when the null hypothesis is true, some circles will be plotted by chance alone. Besag and Newell have further discussion of some possible ways to proceed.

5.4   Fotheringham and Zhan's Method

This method is similar to Openshaw's Geographical Analysis Machine. A circle with a radius chosen randomly within a prespecified range is drawn with its center at a location chosen at random within the study region. Compared with Openshaw's GAM, this method has considerably reduced computational intensity because the size and location of each circle is chosen randomly. The circle is drawn on the map if the number of incidents inside the circle is so sufficiently great that it is unlikely to have occurred by chance. The significance of the circle can be evaluated on the basis of the Poisson probability associated with the observed number of cases in a given population. Fotheringham and Zhan (1996) compare their method with those of Besag and Newell, and Openshaw, and find that "the Besag and Newell method appears to be particularly good at not identifying false positives, although the Fotheringham and Zhan method is easier to apply and is not dependent on a definition of minimum cluster size."

5.5   Cluster Evaluation Permutation Procedure

Turnbull et al. (1990) suggest a method in which the study area is first divided into a large number of regions, and the number of observed cases and the associated population are assigned to each. For each region, the region is combined with the nearest surrounding regions to form a "ball" containing a predefined, fixed number of persons. The outermost region may be fractionally included in the ball to ensure that the ball has the predefined total population. For instance, if the predefined population is 3000, the central zone in Figure 5.2 is to be combined with only two-thirds of the nearest region.

For each ball, the number of incidents that are inside is counted. Because all balls have exactly the same population, those counts can be seen as a direct reflection of disease rates within the balls and thus are comparable with one another. Then the analyst determines whether the ball with the maximum number of cases has a number that exceeds the number of cases one would expect if cases were randomly distributed. A critical value for the expected maximum number of cases can be obtained via Monte Carlo simulation. If a significant cluster is found, one can then go on to determine whether the ball with the second highest number of cases has a number that exceeds the number of cases one would expect if cases occurred at random. It should be noted that results of the cluster evaluation permutation procedure are specific to the predefined population size, as with the Besag and Newell method, which uses the predefined number of cases instead of population. A desirable feature of this method is that it accounts for the multiple testing problem by evaluating only a ball with the maximum number of cases.

5.6 Exploratory Spatial Analysis Approach of Rushton and Lolonis

An exploratory method for finding spatial clusters when data consist of point locations has been suggested by Rushton and Lolonis (1996), in the context of a study of birth defect rates in Des Moines, Iowa.

To begin, they overlay a set of grid points on top of the study area, and then construct circles around each grid point. The number of events (in their case, birth defects) that occur within each of the circles is counted, as well as the number of births within each circle. Based on these counts, a rate of birth defects corresponding to each grid point is computed. The result is displayed as an isarithmic map, with isolines connecting areas with equal rates.

To assess whether the rate observed at any grid point is "unusually" high, simulations are conducted as follows. The location of births is taken as fixed. Then, birth defects are assigned to birth locations at the rate that prevails across the entire study area (in Des Moines, this was 58 per 1000). Next, the birth defect rates for this simulated scenario were computed for the same grid points and circles. This was repeated 1000 times, and the 1000 birth defect rates at each of the grid points were stored. Finally, at each grid point, the actual birth defect rate was compared with the list of 1000 simulated rates; the ranking of the observed rate on the list of simulated rates served as a measure of significance. For example, if the observed rate ranked 950th on a list of simulated rates (ranked from lowest to highest), the value for that location was assigned a value of $950/1000 \times 100 = 95\%$, reflecting the fact that the observed value was higher than 95% of the simulated values. These percentages were displayed as an isarithmic map, with isolines of equal percentage.

Critical choices again include those related to spatial scale. In the Rushton and Lolonis study, grid points were at locations that were 0.5 miles apart, and circles had a radius of 0.4 miles. Some overlap between circles is desirable, if only to ensure that all births are included in a circle. Clearly, the results will vary with the choice of these parameters. Rushton and Lolonis suggest an exploratory analysis, using, for example, a dark isoline to represent locations where the observations exceed the simulations 95% of the time. A more rigorous confirmatory analysis would need to account for the multiple testing that occurs because of the multiple grid points tested. One way to adjust for this would be to first note the maximum rate that is observed on each of the 1000 simulated maps. Then one could compare the maximum rate on the observed map with the distribution of simulated maximum rates.

5.7 Kulldorff's Spatial Scan Statistic with Variable Window Size

Kulldorff and Nagarwalla (1995) introduce the spatial scan statistic; this can be applied to either point data, where exact locations of individual cases are available, or to areal data, where the population and the number of cases are available for each zone. Similar to the methods discussed earlier, the spatial scan statistic considers a set of circles with varying radii, centered on lattice points.

The spatial scan statistic is based on the likelihood ratio associated with the number of events inside and outside of a circular scanning window. The numerator of the ratio is associated with the hypothesis that the rates inside and outside of the window are different, and the denominator of the ratio is associated with the null hypothesis that the rates inside and outside of the window are the same. Likelihood ratios are computed for circular scanning windows of various sizes, which move to scan over space. The most unusual window under the null hypothesis is the one displaying the maximum likelihood ratio. This maximum observed ratio is compared with maximum ratios that are simulated by assuming the null hypothesis to be true; if, for example, the maximum observed ratio is greater than 95% of the simulated maximum ratios, the cluster is said to be significant using $\alpha = 0.05$.

For each circular window, the null hypothesis to be tested is that the probability of being a case in the circle, p, is the same as the probability of being a case in the rest of the study region, q; that is, $p = q$. The alternative hypothesis is $p > q$, implying clustering of cases in the circle. Let z be the zone within a particular circle, c_z and n_z be the number of cases and the population in z, respectively, and C and N be the total number of cases and the total population in the study area, respectively (see Figure 5.3). Then, the likelihood ratio is defined as

$$L(z) = \frac{L_{p>q}}{L_{p=q}} = \frac{p_z^{c_z}(1-p_z)^{n_z-c_z} q_z^{C-c_z}(1-q_z)^{(N-n_z)-(C-c_z)}}{p_0^C(1-p_0)^{N-C}} \qquad (5.2)$$

where $p_0 = C/N$, $p_z = c_z/n_z$, and $q_z = (C - c_z)/(N - n_z)$. If the observed maximum likelihood ratio is found to be significant, one can proceed to examine whether the second, third, etc., maxima are also significant.

Notable features of this method are that (1) it addresses the multiple testing problem by focusing on the maximum of the observed likelihood ratios, and (2) it is not restricted by predefined population or cluster size.

5.7.1 Example 1: Low Birthweight Cases in California, 2000 (Areal Data)

As mentioned earlier, the spatial scan statistic is applicable to both areal data and point data, so that this and following sections present both types

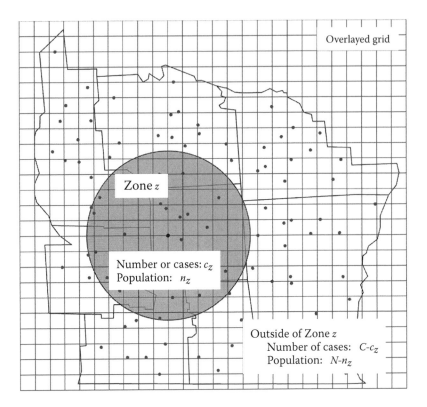

Zone z

Number or cases: c_z
Population: n_z

Outside of Zone z
Number of cases: $C\text{-}c_z$
Population: $N\text{-}n_z$

Overlayed grid

FIGURE 5.3
Spatial scan statistic with a regular lattice basis. Although only one circle is shown here as an example, the method actually examines a large number of circles, as in Figure 5.4.

of examples using the California low birthweight (LBW) data described in subsection 1.7.2.6 and the *SaTScan* software briefly introduced in subsection 1.7.1.4. Here, all records in the dataset are assumed to constitute the risk population, whereas records with low birthweight (i.e., less than 2500 g) are treated as cases.

In this section, let us discuss an areal-data example where LBW cases are aggregated into 58 counties in California. The spatial scan statistic detected two clusters at the significance level of 5% as shown in Figure 5.4a; the triangles in the maps indicate centers of circles circumscribing the clusters.

Cluster 1 is the most likely cluster, with a log likelihood ratio of 18.33 and a *p*-value of .001; it consists of ten counties, including Los Angeles County. The total population and the observed and expected numbers of cases in this cluster are 205,237, 13,110, and 12,610, respectively, resulting in the relative risk of 1.071. Although it is not surprising that Cluster 1 contains Mariposa County, which has the highest LBW rate in the state (see Chapter 1, Figure 1.5), it should also be noted that the cluster contains Mono County, which has a very low LBW rate. This is a potential

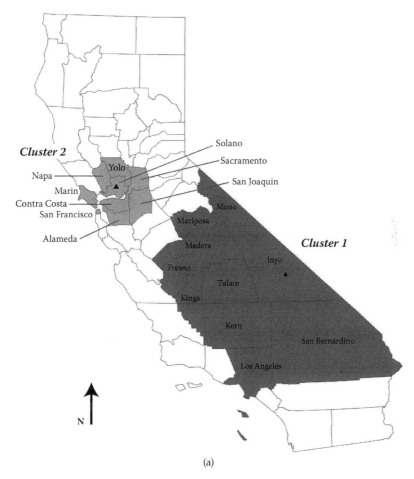

(a)

FIGURE 5.4
Clusters of low birthweight cases in California (2000), detected by the spatial scan statistic (areal data): (a) circular clusters, (b) elliptic clusters.

weakness of the spatial scan statistic or any other method using a scanning window; because a cluster is defined as a set of regions within a given scanning window, it is possible that a region with a low risk is identified as part of the cluster if other regions surrounding it have sufficiently high risks.

The secondary cluster, Cluster 2, consists of nine counties, including San Francisco County, and is geographically much more compact than Cluster 1. The log likelihood ratio and the *p*-value associated with this cluster are 6.43 and .039, respectively. The total population and the observed and expected number of cases are 77,491, 4,982, and 4,761, respectively, and the resulting relative risk is 1.056. It is interesting to note that the log

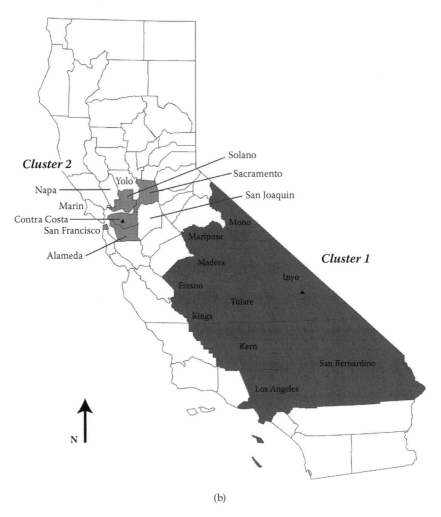

(b)

FIGURE 5.4
(Continued).

likelihood ratio of Cluster 2 is almost one-third that of Cluster 1, whereas
its relative risk is only slightly less than that of Cluster 1.

A natural extension of the spatial scan statistic method is to use scanning
windows of variable shapes that are not limited to circles. Kulldorff et al.
(2006) introduce the elliptic spatial scan statistic, which is a specific case of
the spatial scan statistic that defines the scanning window as ellipses of various sizes, shapes, and angles. They have compared the powers of the circular and elliptic scan statistics using simulated data sets with either circular
or elliptic clusters and conclude that both scan statistics perform compara-
bly well regardless of the true shape of the cluster. A relative advantage of

the elliptic scan statistic is that it may be able to provide a better estimate of the true cluster area. However, they also emphasize the necessity of imposing a mechanism to avoid detecting very narrow, elongated clusters, which are more likely to be artifacts of the method rather than true clusters.

Figure 5.4b shows two significant clusters in the California LBW data, detected by the elliptic spatial scan statistic. Whereas the most likely cluster (Cluster 1) remains the same, the secondary cluster (Cluster 2) becomes smaller, containing only five counties. The five counties in this new Cluster 2 have an overall LBW rate of 0.065 and the four counties that are excluded here have that of 0.060, implying that the elliptic scan statistic has identified this cluster in a more accurate manner than the circular scan statistic. The log likelihood ratio (9.29), the p-value (.010), and the relative risk (1.074) associated with the new Cluster 2 also indicate this.

5.7.2   Example 2: LBW Cases in California (Point Data)

In this section, we discuss an application of the spatial scan statistic to a point pattern data set, where each birth record is represented as a point located at the mother's residential address. As a reminder, there are 482,333 live birth records in this data set and 29,635 records (about 6.1%) are identified as LBW.

The spatial scan statistic detected 21 clusters, among which 7 clusters were significant at the 5% level. The locations and sizes of the 7 significant clusters are shown in Table 5.1 and Figure 5.5.

Table 5.1 shows that the likelihood ratio of Cluster 1 is more than twice that of Cluster 2, whereas relative risk Cluster 2 has a larger. Again, this illustrates that the focus of the spatial scan statistic is not mere concentration of cases relative to the population distribution. It may also be noted that the seven clusters widely vary both in terms of size (i.e., radius) and in terms of the number of cases included, illustrating the second notable feature of the method mentioned earlier.

When compared with the results presented in the previous section, clusters 1, 3, 4, 5, and 7 here seem to correspond with the former Cluster 1, and clusters 2 and 6 correspond with the former Cluster 2. However, the seven clusters detected in the point data are much smaller than the two detected in the areal data, even when combined together, implying large spatial variability in the distribution of LBW cases within each county.

It may be noted that Mariposa County, which has the highest county-level LBW rate, does not contain any significant clusters. Because this county had only 86 live births in 2000, when Los Angeles County, for example, had 146,412 births, its high LBW rate is more likely to be a result of instability associated with the small number rather than a reflection of an actually raised risk. The spatial scan statistic applied to point data is advantageous in such a situation because clusters to be detected are not restricted by predetermined zones.

TABLE 5.1

Clusters of Low Birthweight Cases in California, 2000, Detected by the Spatial Scan Statistics

Cluster ID	Center Latitude	Center Longitude	Radius (km)	Likelihood Ratio	p-Value	Observed # of Cases	Expected # of Cases	Obs./Exp. Ratio	Relative Risk
1	33.90	−118.37	14.11	41.49	.001	1907	1557.47	1.22	1.24
2	38.47	−121.45	3.45	17.48	.001	150	90.75	1.65	1.66
3	34.16	−118.16	5.77	15.51	.004	272	192.99	1.41	1.41
4	34.13	−116.22	45.65	15.23	.004	245	171.05	1.43	1.44
5	33.82	−118.16	3.81	14.94	.007	183	120.79	1.51	1.52
6	37.82	−122.11	32.09	14.23	.015	2347	2114.43	1.11	1.12
7	34.73	−117.35	72.70	13.26	.044	1218	1055.31	1.15	1.16

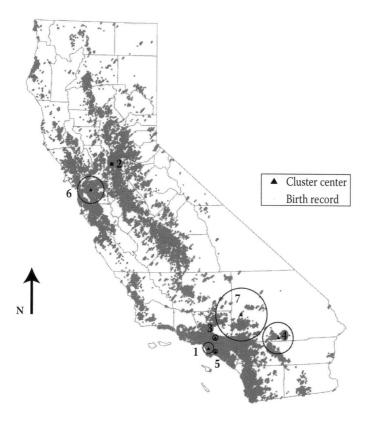

FIGURE 5.5

Clusters of low birthweight cases in California (2000), detected by the spatial scan statistic (point data). (Note: Locations of the birth records were randomly moved to protect confidentiality.)

5.8   Bonferroni and Sidak Adjustments

The most common method of correcting for multiple testing is the Bonferroni adjustment. If m separate, independent tests are made, then instead of choosing a critical value of the test statistic using α as the Type I error probability for each test, one uses α/m for each test. For one-sided tests, a Type I error will be committed when at least one of the m observed local statistics exceeds its critical value (say, b). The overall Type I error is then equal to the probability that the maximum local statistic is greater than b. This in turn is equal to one minus the probability that all m tests accept the null hypothesis, that is,

$$\Pr\left(\max_{i;i=1..,m} z_i > b \right) = 1 - \left(1 - \frac{\alpha}{m} \right)^m \tag{5.3}$$

where z_i is the local statistic for region i. This probability is just slightly less than the desired overall Type I error of α, whenever $m \geq 2$. David (1956) provides an early description and application of the Bonferroni adjustment.

A small improvement is possible by using the Sidak adjustment (e.g., see Sidak 1968), which replaces the α/m that is used for the Bonferroni correction with $1 - (1 - \alpha)^{1/m}$. Then, when tests are independent, $\Pr(\max_{i;i=1..,m} z_i > b) = \alpha$.

The Bonferroni and Sidak adjustments are accurate when the multiple tests are independent of one another, but when the outcomes of individual tests are positively correlated, the adjustment will be conservative. With m positively correlated tests, the "effective" number of independent tests is less than m. For example, if the outcomes of all m tests were perfectly correlated with one another, there would effectively be only one test (because upon knowing the outcome of one test, you would know the outcomes of all other tests). Hence, in this case no adjustment for multiple testing would be necessary; if a Bonferroni or Sidak adjustment was made, the actual Type I error probability would be much less than α, and the power of the test would be reduced. To say that using a Bonferroni or Sidak adjustment for m tests is conservative means that the adjusted critical value is too extreme, significant results may be missed, and the actual Type I error is less than the nominal value of α. The situation is illustrated in Figure 5.6, where CV_0 is the critical value associated with the original significance level α and CV_{adj} is the critical value obtained by either Bonferroni or Sidak adjustment. Although CV_0 is too liberal, implying that it would reject the null hypothesis too often, CV_{adj} is too conservative and would lead to a greater probability of a Type II error, accepting the null hypothesis when it is not true. An appropriate critical value that takes into account both multiple testing and positively correlated tests should be located somewhere between CV_0 and CV_{adj}.

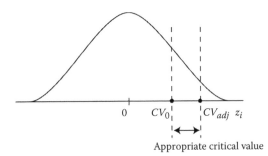

Appropriate critical value

FIGURE 5.6
Critical values with and without adjustment for multiple testing (CV_0: critical value associated with the original significance level α; CV_{adj}: critical value associated with the Bonferroni- or Sidak-adjusted significance level).

Local tests carried out for neighboring geographic areas are positively correlated because some of the same data are used in the computation of their statistics. It is of interest to know just how conservative the Bonferroni and Sidak adjustments are, and whether there are better, alternative adjustments that can be made. These issues are addressed later in this chapter.

5.8.1   Power Loss with the Bonferroni Adjustment

One problem with the Bonferroni method of adjustment, even when tests are indeed independent, is that the resulting test will lack power when a large number of tests are made. To illustrate, suppose we wish to test the null hypothesis that the true mean of data coming from a normal distribution with known standard deviation (equal to one) is equal to zero. If the alternative hypothesis that the data come from a $N(\delta,1)$ distribution is true, and if $\alpha = 0.05$, the power of the usual one-sided z-test is

$$\Pr\left\{z = \bar{x}/(1/\sqrt{n}) > 1.645\right\} = 1 - \Phi\left(1.645 - \delta\sqrt{n}\right) \qquad (5.4)$$

$\Phi(z)$ is the cumulative distribution function for the standard normal distribution, returning the probability that a standard normal variate assumes a value less than z. For example, if $\delta = 0.12$ and the sample size, n, is equal to 40, the power of this single test is 0.188.

Now suppose that there are two separate tests, $m = 2$ (e.g., suppose that we have two regions, and we carry out one of these tests in each region). To maintain the overall Type I error level at $\alpha = 0.05$, we use a Bonferroni adjustment, and this implies that a critical value of Φ^{-1} $(0.05/2) = 1.96$ should be used with each test. The power for a single test is now given by

$$\Pr\left\{z = \bar{x}/(1/\sqrt{n}) > 1.96\right\} = 1 - \Phi(1.96 - \delta\sqrt{n}) \qquad (5.5)$$

and this is also the expected proportion of tests rejected. With $\delta = 0.12$ and $n = 40$, this is equal to 0.115. Note that we are now less able to reject false null hypotheses (i.e., when there are two tests, we correctly reject only 11.5% of the time, whereas we were able to reject the false null hypothesis 18.8% of the time when there was only one test). When there are m separate tests,

$$\Pr\left\{z = \bar{x}/(1/\sqrt{n}) > \Phi^{-1}(\alpha/m)\right\} = 1 - \Phi\left(\Phi^{-1}(\alpha/m) - \delta\sqrt{n}\right) \qquad (5.6)$$

For a given δ, the proportion of tests rejected will decline as the number of tests m increases. With $\delta = 0.12$, $n = 40$, and $m = 100$, the critical value of z is equal to 3.29, and only 0.0057 of the tests are rejected.

The "price" associated with retaining an overall level of Type I error at α is a reduced ability to detect real deviations from the null hypothesis.

This is to be expected; if we really only want a probability α of rejecting *any* of m hypotheses, we have to be conservative in testing each one if m is large. However, it is also possible to improve upon Bonferroni's method, as discussed in the following section.

5.9   Improvements on the Bonferroni Adjustment

Holm (1979) suggested the following sequential procedure to lower the power loss associated with the Bonferroni adjustment. Rank the m p-values, with p_1 being the smallest and p_m the largest. Compare p_1 with α/m; if $p_1 > \alpha/m$, then none of the tests should be rejected. If $p_1 < \alpha/m$, reject that hypothesis and then compare p_2 with $\alpha/(m-1)$ (because there are at most $m - 1$ true null hypotheses). Continue in this manner until for some $j, p_j > \alpha/(m - j + 1)$. The first $j - 1$ hypotheses are rejected and one fails to reject the others. As Aickin and Gensler (1996) note, this is guaranteed to (a) bound the Type I error at α, and (b) reject at least as many hypotheses as the Bonferroni procedure. Whereas the "Bonferronied" p-value is $\min(mp_i, 1)$ for hypothesis i, the "Holmed" p-value is $\min\{\max[(m-j+1)p_j; j = 1, ..., i], 1\}$, and the latter will never be greater than the former. The "Holmed" p-value can be thought of as the smallest overall significance level at which hypothesis i would be rejected.

Hochberg (1988) presents a similar method, but this time the p-values are considered from highest to lowest. First, the largest p, p_m, is compared with α; if $p_m < \alpha$, all hypotheses are rejected. If it is greater, the hypothesis is not rejected, and then p_{m-1} is compared with $\alpha/2$. One continues sequentially until for some $j, p_j < \alpha/(m - j + 1)$. The first $j - 1$ hypotheses considered are accepted and the remainder are rejected.

A graphical method based on Holm's concept is provided by Schweder and Spjotvoll (1982). They recommend plotting N_p, the number of p-values greater than p, versus $1 - p$ as shown in Figure 5.7. If all of the null hypotheses are true, the plot will follow a straight line, with slope equal to the number of tests, m. If some of the hypotheses are not true, the plot will deviate from a linear fit on the right-hand side of the graph, and in particular, points will lie above the straight line of best fit. These outlying points correspond to hypotheses that have small p-values, that is, those to be rejected. The authors recommend fitting a line to the linear portion of the graph, and then estimating the number of true null hypotheses as the slope of the line (S), which will be less than m. They suggest a formal test using a Bonferroni adjustment based upon this slope S, instead of the usual m. Thus, a critical z-score would be found from $\Phi^{-1}(\alpha/S)$ instead of $\Phi^{-1}(\alpha/m)$. Both Schweder and Spjpotvoll (1982) and Haybittle et al. (1995) find this still too conservative, and recommend rejecting those hypotheses

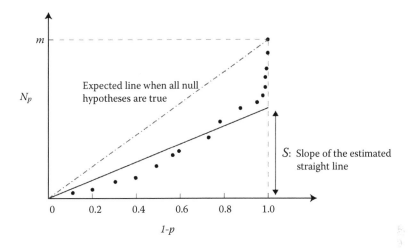

FIGURE 5.7
Graphical evaluation of multiple p-values proposed by Schweder and Spjotvoll (1982).

corresponding to all of the points that deviate from the straight line in the right-hand side of the graph. Although this may be of value in uncovering more false null hypotheses, one should be cautioned that the Type I error is no longer bounded by α and, hence, this strategy should be considered more exploratory.

Hochberg and Benjamini (1990) modify Schweder and Spjotvoll's graphical procedure and combine it with the Holm and Hochberg methods. Their graphical procedure will improve upon the Holm and Hochberg methods, but the degree to which the Type I error is bounded by α will depend on how well the number of true hypotheses is estimated by the regression line. Brown and Russell (1997) provide a comparison of various methods, and further details on these and other procedures are given by Westfall and Young (1993).

5.10 Rogerson's Statistical Method for the Detection of Geographic Clustering

The last method discussed in this chapter was proposed by Rogerson (2001a) and is founded on kernel-based, smoothed estimates of spatial variables of interest, which makes this method distinguishable from those presented earlier. Although kernel-based smoothing is a way to represent the spatial variability in the mean of the variable, its use has been mostly for visual, exploratory analysis, and attempts to assess the significance of

the peak of the smoothed surface have been limited to Monte Carlo simulation (e.g., Kelsall and Diggle 1995) or to more formal statistical methods that do not control properly for the likelihood of a Type I error (e.g.,
Bowman and Azzalini 1997). Rogerson's approach is based on Worsley's
(1996) work on the maxima of Gaussian random fields. This provides a
way to assess the significance of the maximum of a set of local statistics
obtained as kernel-based, smoothed surface values without resorting to
Monte Carlo simulation methods.

The method may be summarized as follows. For a study area consisting
of a set of regions, each with unit area, the method first constructs local
statistics z_i for each region using the standardized original observations
(denoted y_i), and weights defined as $w_{ij} = (\sqrt{\pi}\sigma)^{-1}\exp(-d_{ij}^2/2\sigma^2)$, where
d_{ij} is the distance from region i to region j and σ is chosen as the standard
deviation of a normal distribution that matches the size of the hypothesized cluster. When the study area consists of a regular lattice of square
cells and a guard area is defined at its edges, the local statistic for region i
is defined as $z_i = \sum_j w_{ij}y_j$. The local statistics constructed at or near edges
will not have as many regional neighbors as other regions. Consequently,
the sum of the squared weights ($\sum_j w_{ij}^2$), and the variance of the local statistic (which is based on the sum of the squared weights), will be smaller
for regions near edges than for other regions. To help address these edge
effects, as well as similar effects induced by regional configurations other
than a square grid, it is useful to use modified, scaled weights in place of
the original weights to ensure equal variances for all local statistics. The
modified weights are defined in terms of the original weights as follows:

$$w_{ij}^* = \frac{w_{ij}}{\sqrt{\sum_j w_{ij}^2}} \tag{5.7}$$

The local statistics, z_i, all have normal distributions with mean zero and
variance one (e.g., see Siegmund and Worsley 1995; Rogerson 2001a).

The next step is to find the critical value M^* such that $\Pr(M = \max z_i > M^*)$
$= \alpha$ by using that value of M^* which leaves probability $(1 + .81\sigma^2)\alpha/A$ in the
tail of the standard normal distribution, for the study area that has been
subdivided into a grid of A square cells, each having side of unit length.
Alternatively, M^* may be approximated by

$$M^* = \sqrt{-\sqrt{\pi}\ln\left(\frac{4\alpha(1+.81\sigma^2)}{A}\right)} \tag{5.8}$$

The more general case, where regions have not been scaled to unit area, is
summarized in Chapter 11.

In the next subsection we explore the foundations of this approach in a bit
more detail.

5.10.1   The Geometry of Random Fields

Critical values for the maximum among a set of local statistics may be based on the geometry of random fields. Let $\mathbf{x}$ be a location in d-dimensional space, and let $\mathbf{Y}(\mathbf{x})$ denote a random multivariate value observed at $\mathbf{x}$. A random field is defined by the set of values $\mathbf{Y}(\mathbf{x})$ for some subset of interest within the d-dimensional space (Cressie 1993). Here we will confine our attention to univariate random fields in $d = 2$ dimensions, though results are also available for cases where the number of dimensions is other than 2. We will also pay particular attention to the special case of a Gaussian random field, where the values at each location are taken from a Gaussian distribution. Results for other types of random fields, including χ^2, t, and F fields, are also available (e.g., see Worsley 1994).

Recent developments have improved upon and generalized the pioneering work of Adler (1981), who derived an approximation for the probability that the maximum of a Gaussian random field would exceed a specified value. In particular, Worsley (1994) has used principles of integral geometry to derive the following, improved version of Adler's original expression for exceedance probabilities. In two dimensions, for the case where independent observations are observed at many points on a lattice, and then smoothed using a Gaussian kernel, it is:

$$\Pr\left(M = \max_i z_i > z^*\right) = \frac{AM^*\varphi(M^*)}{4\pi\sigma^2} + \frac{D\varphi(M^*)}{\sqrt{\pi}\sigma} + [1 - \Phi(M^*)] \qquad (5.9)$$

where $\varphi(\cdot)$ and $\Phi(\cdot)$ are, respectively, the probability density and cumulative distribution function of a standard normal variate. D denotes the caliper diameter and A the area of the study region. The caliper diameter is the average of the diameter as measured through all rotations of the study area. For a rectangle with sides a and b, the caliper diameter is $(a + b)/2$; for a circular study region of radius r, the caliper diameter is equal to $2r$. Again, Equation 5.9 gives the probability that the maximum of a Gaussian random field, when smoothed by a Gaussian kernel, exceeds z^*. Rogerson (2001a) shows how Equation 5.9 may be modified and solved approximately to arrive at the direct expression for the critical value given in Equation 5.8.

5.10.2   Illustration

A 30×30 grid was filled with y-values generated from a $N(0,1)$ distribution. For this simulation of the null hypothesis of no local cluster, values of $\sigma = 1, 2$, and 3 were used with the Gaussian kernel to smooth the initial y values, creating a 30×30 grid of local statistics, z_i. To avoid edge effects, the 22×22 grid occupying the center of the 30×30 grid was searched for the maximum z_i value.

TABLE 5.2

Simulated and Approximate Critical Values ($\alpha = 0.05$) for the Maximum Local Statistic When $A = 484$

(1) σ	(2) Simulated 95th Percentile	(3) Critical Value By Equation 5.9 (Type I Error Probability)	(4) Critical Value Adjusted for Discreteness (Type I Error Probability)	(5) Equation 5.16	(6) Equation 5.17
1	3.575	3.779 (.029)	3.556 (.052)	3.556	3.572
2	3.342	3.389 (.043)	3.311 (.053)	3.328	3.354
3	3.110	3.150 (.041)	3.112 (.050)	3.136	3.172

Using the values of $A = 484$ and $D = 22$ in Equation 5.9, and setting the left-hand side equal to $\alpha = 0.05$ yields critical z-values of 3.779, 3.389, and 3.150 for the cases where $\sigma = 1, 2$, and 3, respectively. For comparison purposes, the 95th percentiles for maximum z_i values were then found from 1000 Monte Carlo simulations.

Results are shown in the first three columns of Table 5.2, along with the Type I error probabilities associated with using the critical value derived from Equation 5.9. Although the Type I error probabilities are close to their nominal value of .05 for the latter two cases, use of Equation 5.9 would be overly conservative in the case where $\sigma = 1$. In fact, when $\sigma = 1$, Equation 5.9 is even more conservative than a Bonferroni adjustment, because with $A = 484$ cells, the value of M^* following a Bonferroni adjustment is 3.711.

Equation 5.9 is based on the assumption of a continuous random field, and its poor performance for the case $\sigma = 1$ is due to the discreteness of the grid used to generate the observed values.

5.10.3   Approximation for Discreteness of Observations

Adjusted critical values may be found by first determining the amount of smoothing implicit in the initial discrete grid, which represents a set of aggregated or smoothed observations. We can represent the initial data as a smoothed Gaussian field in the following way. With $A = 484$, the critical value associated with a Bonferroni adjustment is 3.711. By using $zM^* = 3.711$ in Equation 5.9, we can solve for the amount of smoothing that is imparted by the square grid ($\sigma = \sigma_0$). In particular, Equation 5.9 may be rearranged so that σ_0 is the solution to the following quadratic equation:

$$(\alpha - 1 + \Phi(M^*))\sigma_0^2 - \frac{D\varphi(M^*)}{\sqrt{\pi}}\sigma_0 - \frac{AM^*\varphi(M^*)}{4\pi} = 0 \qquad (5.10)$$

In our example, $A = 484$, $D = 22$, $\alpha = 0.05$, $M^* = 3.711$, and solving for σ_0 yields $\sigma_0 = 1.133$. In subsection 5.10.5, an argument is presented that suggests that for most problems, this step is not necessary, and the value of σ_0 may always be taken as $10/9 = 1.111$.

The total amount of smoothing in each case (σ_t) is then defined by combining the implied initial smoothing brought about by the discreteness of the grid (σ_0) with the smoothness chosen for the Gaussian kernel in defining the local statistics (say, σ_l). Thus

$$\sigma_t = \sqrt{\sigma_0^2 + \sigma_l^2} \tag{5.11}$$

These choices imply respective σ_t values of 1.495, 2.288, and 3.199 for the cases where $\sigma_l = 1$, 2, and 3, respectively. Using these values in Equation 5.9 and setting the left-hand side equal to $\alpha = 0.05$ results in the critical values of $M^* = 3.556, 3.311$, and 3.112 in the $\sigma_l = 1$, 2, and 3 cases, respectively. These values are shown in column 4 of Table 5.2; their associated p-values (i.e., Type I error probabilities associated with these critical values) are in close agreement with the nominal value of .05.

5.10.4 Approximations for the Exceedance Probability

Of the terms on the right-hand side of Equation 5.9, the first term contributes most to the p-value on the left-hand side. Table 5.3 reveals the contributions of each term on the right-hand side of Equation 5.9 to the nominal Type I error probability of 0.05 for the illustration used in Sections 5.10.2 and 5.10.3, which now includes a correction for the discrete number of spatial units.

The table shows that the first term is by far the most important, and in each case the two-term approximation

$$\Pr\left(M = \max_i z_i > M^*\right) \approx \frac{AM^*\varphi(M^*)}{4\pi\sigma^2} + \frac{D\varphi(M^*)}{\sqrt{\pi}\sigma} = \frac{\varphi(M^*)(AM^* + 4\sqrt{\pi}D\sigma)}{4\pi\sigma^2}$$

$$\tag{5.12}$$

TABLE 5.3

Contribution of Terms in Equation 5.9 to the Type I Error Probability

σ_t	First Term	Second Term	Third Term
1.495	.0439	.0059	.0002
2.288	.0405	.0090	.0005
3.199	.0369	.0122	.0009

TABLE 5.4

Approximations for the Critical Value M^*

(1) $\sigma/\sqrt{A}$	(2) M^* (Equation 5.9)	(3) M^* (Equation 5.12) Two-Term Approximation	(4) M^* (Equation 5.13) One-term Approximation	(5) Resels Equation 5.17 ($k = 0.9$)
0.01	4.535	4.535 (.050)	4.532 (.051)	4.463 (.068)
0.02	4.205	4.205 (.050)	4.196 (.052)	4.160 (.060)
0.05	3.727	3.727 (.050)	3.700 (.055)	3.716 (.052)
0.10	3.334	3.331 (.050)	3.267 (.061)	3.349 (.047)
0.15	3.094	3.087 (.050)	2.977 (.069)	3.118 (.047)
0.20	2.922	2.909 (.052)	2.748 (.079)	2.944 (.047)
0.25	2.790	2.768 (.053)	2.552 (.090)	2.803 (.048)

should suffice (because the sum of the first two terms is close to 0.05). When the amount of smoothing imparted by the kernel is sufficiently small, the one-term approximation

$$\Pr(\max_i z_i > M^*) \approx \frac{AM^* \varphi(M^*)}{4\pi\sigma^2} \tag{5.13}$$

will be adequate.

The use of the approximations (Equations 5.12 and 5.13) will result in critical values of M^* that are slightly lower than those derived with the full, three-term expression in Equation 5.9. Table 5.4 reveals that if $\sigma/\sqrt{A}$ is greater than about 0.05 or 0.10, the one-term approximation in Equation 5.13 will be too liberal, and should be abandoned in favor of the approximation in Equation 5.12.

5.10.5 An Approach Based on the Effective Number of Independent Resels

These approximations still require a numerical solution for the desired critical value, M^*, and it is of interest to ask whether a simpler solution for M^* is possible. One possibility is to attempt an estimate of the effective number of spatial units, or *resels*, upon which to base a Bonferroni adjustment.* The greater the amount of smoothing, the less accurate will be the Bonferroni adjustment, which is based on all A cells and, hence, we seek a value for the number of resels, r, that will be less than A. Let us take

$$r = \frac{A}{(k\sigma)^2} \tag{5.14}$$

---

* This definition of resels is similar in concept, though different in detail, when compared with that used by Worsley et al. (1992).

TABLE 5.5

The Relative Flatness of k as a Function of $\sigma/\sqrt{A}$

$\sigma/\sqrt{A}$	M^*	k
.01	4.535	0.76
.02	4.205	0.81
.05	3.727	0.88
.10	3.334	0.92
.15	3.094	0.94
.20	2.922	0.93
.25	2.790	0.92
.30	2.683	0.90
.35	2.596	0.88
.40	2.523	0.85

where k is an empirical constant of proportionality. For a grid of square cells, the idea here is to divide the study area into a number of resels that is directly proportional to the number of cells, and inversely proportional to the amount of smoothing, as measured by the variance σ of the Gaussian kernel.

A simple Bonferroni adjustment based on r turns out to be possible only because the value of k is approximately constant throughout a wide range of $\sigma/\sqrt{A}$ values. To illustrate, the value of k satisfying

$$\Phi^{-1}(1-\alpha/r) = \Phi^{-1}(1-\alpha(k\sigma)^2/A) = M^* \qquad (5.15)$$

was determined for each of the rows in Table 5.5, using $\sigma = 0.05$, and using the values of $\sigma/\sqrt{A}$ and M^* given in each row (where the value of z^* is determined from Equation 5.9). Table 5.5 shows that the value of k, as a function of $\sigma/\sqrt{A}$, is relatively flat over much of its range. This suggests that a very good approximation for M^* may be found by taking $k = 0.9$, and therefore setting the number of resels equal to $r = A/(.9\sigma)^2$. A Bonferroni adjustment based on this number of resels then yields the desired critical value, M^*:

$$M^* \approx \Phi^{-1}\left(1 - \frac{\alpha(.9\sigma)^2}{A}\right) \qquad (5.16)$$

Thus, one can determine the approximate critical value by finding the z-value that leaves $\alpha(.9\sigma)^2/A$ in the tail of the standard normal distribution. Because this may require the use of a detailed z-table that provides areas for relatively high z-values, it is also of interest to find an approximation that does not require the use of such a detailed z-table. Using tight bounds for the cumulative distribution function of a normal variable

(Sasvari 1999), m^* may also be approximated by

$$M^* \approx \sqrt{-\sqrt{\pi}\ln\left(\frac{4\alpha}{r}\right)} = \sqrt{-\sqrt{\pi}\ln\left(\frac{4\alpha(1+.81\sigma^2)}{A}\right)} \qquad (5.17)$$

Note from Table 5.5 that as $\sigma/\sqrt{A}$ declines below about 0.02, these approximations will not work as well because the value of k begins to decline away from 0.9. However, as column 5 of Table 5.4 shows, the use of $k = 0.9$ when $\sigma/\sqrt{A}$ is as small as 0.01 results in critical values that are only slightly liberal.

Columns 5 and 6 of Table 5.2 demonstrate the adequacy of the approximations given by Equations 5.16 and 5.17 for the case of the maximum local statistic observed on the 22 × 22 grid.

The value of $k = 0.9$ suggests that we can use $\sigma_0 = 1/0.9 = 1.111$ as a measure of the smoothing implicit in the discrete grid. This is because when there is no additional smoothing (i.e., local statistics are based only on the value in the local cell, and weights associated with surrounding cells are zero), the local statistics in each cell are independent, and $r = A$. Using this in Equation 5.14 with $k = 0.9$ requires a standard deviation of $\sigma_t = \sigma_0 = 1/0.9 = 1.111$.

If the amount of smoothing (σ) is greater than or equal to one, the fact that the resel approach can be used when $\sigma/\sqrt{A} > .01$ implies that the approach may be adopted when $A < 10,000$. If the grid is finer than $100 \times 100 = 10,000$ or $\sigma < 1$, then one should ensure that the total amount of smoothing yields $\sigma_t/\sqrt{A} > 0.01$ before proceeding with the critical z-value based on resels.

An approximate p-value corresponding to the observed test statistic, $M = \max z_i$ may be found by rearranging the expression for the critical value, M^*. Specifically, the expression is solved for α; when M^* is replaced by the observed test statistic (M), α may be interpreted as the p-value:

$$p = \frac{A}{4(1+0.81\sigma^2)}e^{-(M^2/\sqrt{\pi})}$$

This approximation for the p-value is accurate for small p-values.

Finally, when a study region consists of subregions that are other than regular, square cells, the appropriate threshold value varies over space because the amount of smoothing and thus the effective number of spatial units is not spatially constant. Consequently, "local" threshold values would be technically necessary to appropriately evaluate the maximum local statistics. An alternative approach would be to use adaptive smoothing, where the standard deviation associated with the weights varies over space.

5.10.6   Example

Mercer and Hall (1911) give the wheat yields for a 20 × 25 grid of plots covering an area of approximately one acre in Rothamsted, England. The

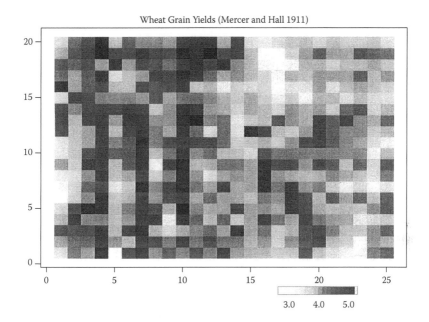

FIGURE 5.8
Spatial distribution of standardized yields.

data are provided in Cressie (1993), and Figure 5.8 depicts the geographical distribution of standardized yields. We seek to determine the presence and location of any subregions of significantly high yield.

With $\sigma_l = 1$, $\sigma_t = \sqrt{1.111^2 + 1^2} = 1.495$. The map of local statistics is displayed in Figure 5.9. Among the $16 \times 21 = 336$ cells of the grid's interior (implying a guard area of width equal to two cells, along all edges of the study area), the observed maximum of 3.502 is found for the local statistic in row 18, column 11 of the original data. This is just slightly greater than the critical value of 3.461 found using Equation 5.16 and the critical value of 3.480 found using Equation 5.17.

With $\sigma_l = 2$, $\sigma_t = \sqrt{1.111^2 + 2^2} = 2.288$. If attention is confined to the 12×17 grid of 204 interior cells to avoid possible edge effects (note that the guard area is chosen to be larger than for the case where $\sigma_l = 1$ because the kernel width is also larger), the maximum observed statistic is 3.436, observed for the cell at row 12, column 6 of the original data. This exceeds the respective critical values from Equations 5.16 and 5.17 of 3.079 and 3.117. Figure 5.10 depicts the location of significantly high wheat yields; the figure was produced by first constructing an interpolated surface of the local statistics, and then displaying only that portion of the surface that exceeded the critical value of 3.079.

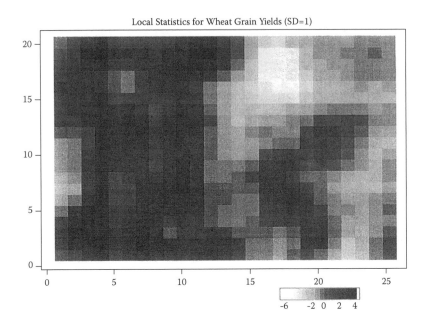

FIGURE 5.9
Map of local statistics; σ = 1.

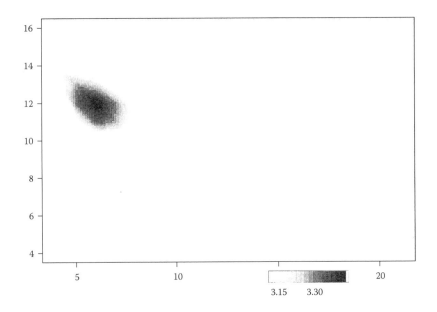

FIGURE 5.10
Regions with significant local statistics; σ = 2.

5.10.7   Discussion

In this section, we have considered a local statistic based on a Gaussian kernel. The statistic has the desirable feature that one may easily find the critical value (via Equations 5.16 or 5.17) necessary for testing the significance of the maximum of the local statistics defined over a study area. The statistic relies on the assumption that the underlying data come from a normal distribution. In choosing kernels for smoothing, it is commonly noted that the choice of a kernel function is much less important than the choice of the bandwidth. Because estimates are relatively robust with respect to the form of the kernel function, the Gaussian kernel is a good choice when one is interested in assessing the significance of maxima, as it lends itself readily to such testing. In addition, one should be aware that the choice of bandwidth should be made to match the hypothesized cluster size; in the different context of optimal estimation of kernel surfaces, bandwidth choice could be quite different.

The distribution of local statistics is also affected by global spatial autocorrelation. In particular, the presence of global spatial autocorrelation will make it more difficult to detect significant local statistics. Recent developments in the study of random fields (e.g., Worsley et al. 1999) suggest that the aforementioned approach described might also be modified to find an approximate critical value associated with the maximum local statistic in the presence of global spatial autocorrelation.

Finally, there are situations in which one may be interested in trying different amounts of smoothing (i.e., choose various values for σ) to see at which scale local statistics are most significant. Kulldorff's spatial scan statistic handles this case using a Monte Carlo approach. Siegmund and Worsley (1995) provide details on how critical values may be derived analytically when one wishes to test a range of σ-values.

6

Retrospective Detection of Changing Spatial Patterns

6.1 Introduction

Until this point we have focused upon the statistical testing of the null hypothesis of no geographic clustering. Time has not entered our analyses in any way. Data for a single time period have been used to assess the null hypothesis during that period.

However, it is often desirable to detect whether change over time has occurred in a set of data. In an aspatial context, we may wish to know whether the mean of a variable has increased over time. In a spatial context, interest may lie in questions of whether points that are close together in space are also close together in time, or whether the spatial pattern of observations has changed at any time in the past. In this chapter, the tests we examine are retrospective, in the sense that just one test is carried out on a given set of data.

In Section 6.2, we review the Knox test—a classic test for space–time interaction. After that, we focus on tests that are designed to test the null hypothesis of no change against the alternative that there is a single *change-point*, where either the mean or a multivariate set of probabilities changes. We will discuss and illustrate two tests: in Section 6.3 we describe a test for a single change in the mean of a series of normally distributed variables, and in Section 6.4 we describe a test for a change in multinomial distribution. This latter test will be particularly useful for spatial analysts when the multinomial categories are defined as geographic regions, in comparison with the former one, which is essentially aspatial.

6.2 The Knox Statistic for Space–Time Interaction

Knox (1964) proposed a test of space–time interaction based upon a count of the number of event pairs that occur within a prespecified, critical interval of time (say, t_0), and a prespecified, critical distance (say, s_0). With n points located in space and time, there are $n(n - 1)/2$ distinct pairs of

TABLE 6.1

The Knox Test for Space–Time Interaction

	Close in Space	Not Close in Space	Total
Close in time	n_{st}	$n_t - n_{st}$	n_t
Not close in time	$n_s - n_{st}$	$n(n-1)/2 - n_s - n_t + n_{st}$	$n(n-1)/2 - n_t$
Total	n_s	$n(n-1)/2 - n_s$	$n(n-1)/2$

points. These pairs may be divided into the two-by-two matrix shown in Table 6.1, which indicates whether pairs are close together in space or close together in time, or both.

Let n_s be the observed number of pairs that are close together in space (i.e., pairs that are separated by a distance less than s_0). Let n_t be the observed number of pairs that are close together in time (i.e., pairs that are separated in time by less than t_0). The test statistic is simply n_{st}, the observed number of pairs that are close in both space and time. The test statistic will exceed its expectation of $\frac{2 n_s n_t}{n(n-1)}$ when point pairs that are close together in space are also closer than expected in time (or, alternatively stated, when points that are close together in time are closer than expected in space).

Regarding the distribution of the test statistic, Knox (1964) suggested that the random variable N_{st} (which takes on the observed values n_{st}) was approximately Poisson distributed. Barton and David (1966) showed more precisely that the variance of N_{st} was equal to

$$V[N_{st}] = \frac{2 n_s n_t}{n(n-1)} + 4 \frac{n_{2s} n_{2t}}{n(n-1)(n-2)}$$

$$+ \frac{4\{n_s(n_s-1) - n_{2s}\}\{n_t(n_t-1-n_{2t})\}}{n(n-1)(n-2)(n-3)} - \left\{\frac{2 n_s n_t}{n(n-1)}\right\}^2$$

(6.1)

where

$$n_{2s} = \frac{\sum_i p_{is}^2}{2} - n_s$$

$$n_{2t} = \frac{\sum_i p_{it}^2}{2} - n_t$$

(6.2)

and p_{it} and p_{is} are, respectively, the number of points that point i is closely linked to in time and in space.

Kulldorff and Hjalmers (1999) note that the following alternatives are typically used to carry out Knox's test: (1) assume N_{st} is Poisson, with mean equal to its expectation given above, (2) use the normal approximation to

TABLE 6.2

An Example for the Knox Test ($n = 30$)

	Close in Time	Not Close in Time	Total
Close in space	142	108	250
Not close in space	93	92	185
Total	235	200	435

the Poisson, with mean and variance equal to the expectation, (3) use a normal approximation with mean as mentioned, and variance equal to that given by Barton and David, and (4) carry out a Monte Carlo test by randomly permuting the times among the fixed spatial locations.

6.2.1 Illustration

Let us consider the example shown in Table 6.2, where the total number of points n is 30. The number of distinct pairs is 435 ($= 30 \times 29/2$), and the expected number of pairs that are close both in time and in space is 135.06 ($= 250 \times 235/435$). Based on alternative (1), the probability of observing n_{st} greater than 142 is .258 so that the null hypothesis of no space–time interaction is not rejected at the significance level of 0.05. If we use alternative (2), the observed number of points that are close both in time and in space $n_{st} = 142$ is converted into a z-score of 0.597, which does not exceed the usual (one-sided) critical value of 1.64 so that the null hypothesis is not rejected again. In addition, the one-sided p-value associated with $z = 0.597$ is 0.275, which is similar to the value of $p = .258$ found for the previous approach. Note that alternatives (3) and (4) require more information than shown in the simple summary shown in Table 6.2.

6.3 Test for a Change in Mean for a Series of Normally Distributed Observations

Let x_i be a random variable from a normal distribution with mean μ_i and variance σ^2 observed over time ($i = 1, \dots, n$). Worsley (1979) considers the null hypothesis

$$H_0: \mu_i = \mu; \quad i = 1, \dots, n \tag{6.3}$$

versus the alternative

$$H_1: \quad \mu_i = \mu; \quad i = 1, \dots, k$$
$$\mu_i = \mu'; \quad i = k+1, \dots, n \tag{6.4}$$

The test is thus a retrospective test, where one is interested in the possibility that the mean has changed at time k. It is assumed that μ, μ', k and σ^2 are all unknown.

Worsley gives the statistic for the likelihood ratio test as

$$W = \max_{1 \leq k \leq n-1} \sqrt{n-2} \, |T_k| / S_k \tag{6.5}$$

where

$$T_k^2 = \{k(n-k)/n\}(\bar{X}_k - \bar{X}_{k'})^2$$

$$S_k^2 = \sum_{i=1}^{k} (X_i - \bar{X}_k)^2 + \sum_{i=k+1}^{n} (X_i - \bar{X}_{k'})^2 \tag{6.6}$$

and where $\bar{X}_k$ and $\bar{X}_{k'}$ denote the sample means before and after the change, respectively.

The significance of W may be assessed by comparing the observed value of the statistic with critical values; Table 6.3 is based upon Worsley (1979).

6.3.1   Example

In this example, we use data on observed and expected numbers of breast cancer mortality cases in Ulster County, New York, for the $n = 31$-year

TABLE 6.3

Critical Values of W

n	$\alpha = 0.10$	$\alpha = 0.05$	$\alpha = 0.01$
3	12.71	25.45	127.32
4	5.34	7.65	17.28
5	4.18	5.39	9.46
6	3.73	4.60	7.17
7	3.48	4.20	6.14
8	3.32	3.95	5.56
9	3.21	3.78	5.19
10	3.14	3.66	4.93
15	2.97	3.36	4.32
20	2.90	3.28	4.13
25	2.89	3.23	3.94
30	2.86	3.19	3.86
35	2.88	3.21	3.87
40	2.88	3.17	3.77
45	2.86	3.18	3.79
50	2.87	3.16	3.79

Source: Based upon Worsley, K. J. (1979).

TABLE 6.4

Observed and Expected Cases of Breast
Cancer Mortality in Ulster County,
New York, 1968–1998

Year	Observed	Expected
1968	28	20.95
1969	19	21.18
1970	30	21.88
1971	24	22.48
1972	24	23.56
1973	34	24.25
1974	33	24.42
1975	18	24.42
1976	36	25.17
1977	24	26.04
1978	29	25.80
1979	19	25.84
1980	34	26.60
1981	14	26.75
1982	29	27.15
1983	41	27.29
1984	34	28.21
1985	40	28.36
1986	36	28.31
1987	22	28.60
1988	31	29.28
1989	32	29.61
1990	43	29.87
1991	32	29.73
1992	32	29.27
1993	38	29.46
1994	45	29.09
1995	42	28.99
1996	38	28.26
1997	41	27.36
1998	40	27.02

period 1968–1998 (Table 6.4). The first step is to create normalized z-scores by using the Rossi et al. (1999) transformation

$$z = \frac{O - 3E + 2\sqrt{OE}}{2\sqrt{E}} \tag{6.7}$$

where O and E represent the observed and expected numbers, respectively. The transformed z-scores are shown in Figure 6.1. Treating this transformed

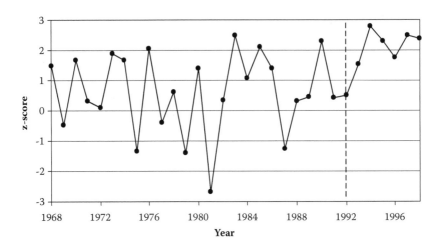

FIGURE 6.1
Rossi-transformed z-scores of the breast cancer mortality data in Table 6.4.

quantity as a normal variable, we find $W = 2.91$, using Equation 6.5, whereas the critical value with $\alpha = 0.10$ is about 2.86 (Table 6.3). The maximum occurs for $k = 25$, implying that the average z-scores observed for the 25-year period 1968–1992 are significantly different from the average z-score observed during 1993–1998. The null hypothesis is rejected at $\alpha = 0.10$. Furthermore, the average z-score before the change is 0.60, and after the change it is 2.21. For the period 1968–1992, there were 738 deaths observed, and 655 expected (obs./exp. = 1.13); for the period 1993–1998, there were 244 deaths observed and 170 expected (obs./exp. = 1.44). Notice, too, that there are no negative z-scores after the changepoint of 1992.

6.4 Retrospective Detection of Change in Multinomial Probabilities

We next turn to the question of retrospective detection of significant changes in spatial patterns. For each time period (e.g., year), multinomial probabilities (p_j) may be defined as the likelihoods that a given case is located in region j. Of interest is a test of the null hypothesis that these multinomial probabilities do not vary over time.

A test of this null hypothesis versus the alternative that there is a single changepoint dividing the sequence of multinomial observations into two distinct subsets ("before" and "after" the changepoint) has been suggested by Srivastava and Worsley (1986). They developed the following test for detecting change in a sequence of multinomial observations.

Data are arranged in an $n \times (m + 1)$ contingency table with ordered rows, where n is the number of time periods and m is the number of regions in the study area minus one, respectively. In this case the rows are ordered by time; although the direction of the ordering is arbitrary, we will adopt the convention of having the most recent information in the last row of the table.

We wish to test for a change in the row proportions after an unknown row, r. Let Q^2 be the usual Pearson χ^2 statistic for testing association between rows and columns in the full table. Let Q_r^2 be the Pearson χ^2 statistic for testing association between rows and columns in the $2 \times (m + 1)$ table formed by aggregating the first $1, \ldots, r$ rows and the remainder of the rows $(r + 1, \ldots, n)$. The test statistic is $Q_{\hat{r}}^2 = \max_r Q_r^2$. Conceptually, we are seeking the aggregation into a table with two rows (a "before change" row, and an "after change" row) that are as dissimilar as possible. Extramultinomial variation can be accommodated; such variation may arise in those situations where the multinomial trials are correlated or where the multinomial probabilities are themselves not known with certainty. The variance inflation factor σ^2 may be estimated as

$$\sigma_{\hat{r}}^2 = \left(Q^2 - Q_{\hat{r}}^2\right)/\{(n-2)m\} \tag{6.8}$$

The test statistic, adjusted for this extramultinomial variation, is $K_{\hat{r}}^2 = Q_{\hat{r}}^2/\sigma_{\hat{r}}^2$, and the quantity $K_{\hat{r}}^2/m$ has an F distribution with m and $(n-2)m$ degrees of freedom.

6.4.1 Illustration

Suppose the number of cases in a three-region system is observed over four time periods; $m = 2$ and $n = 4$ for this data set. Assume no regional change in the size of the at-risk population. The data are as follows:

Time	Region 1	Region 2	Region 3
1	3	5	7
2	5	10	15
3	10	10	10
4	15	10	5

We wish to test the null hypothesis that the set of regional multinomial probabilities does not change over time against the alternative that there is a changepoint where these multinomial probabilities change from one set to another. Note that there are relatively more cases observed in region 3 during the early time periods, but by the last period, relatively more cases observed in region 1. It therefore appears that there has been a shift in the pattern, but perhaps this could simply be a function of random variation.

One quantity that will be needed is the chi-square statistic for association in the entire table. This is found by first finding expected values, which in turn are found by dividing the product of row and column sums by the overall total. For instance, for Region 1 at Time 1, the expected value is computed as $(3 + 5 + 7) \times (3 + 5 + 10 + 15) \div (3 + 5 + 7 + \cdots + 15 + 10 + 5)$. This leads to the following table of expected frequencies:

Time	Region 1	Region 2	Region 3
1	4.71	5	5.29
2	9.43	10	10.57
3	9.43	10	10.57
4	9.43	10	10.57

The chi-squared goodness-of-fit statistic for the overall table, based on the comparison of the observed and expected frequencies, is $Q^2 = 11.41$. Next, we form all possible 2×3 tables, based upon aggregation of row information before and after potential changepoints. For each of these, we find the corresponding expected frequencies:

After $t = 1$:

Observed			Expected		
3	5	7	4.7	5	5.3
30	30	30	28.3	30	31.7

After $t = 2$:

Observed			Expected		
8	15	22	14.1	15	15.9
25	20	15	18.9	20	21.1

After $t = 3$:

Observed			Expected		
18	25	32	23.6	25	26.4
15	10	5	9.4	10	10.6

The chi-square statistics derived from comparing observed and expected frequencies in the usual way are, for each of these three tables, $Q_1^2 = 1.376$, $Q_2^2 = 8.83$, and $Q_3^2 = 8.72$. To assess the significance of the maximum value of 8.83, we first calculate the quantities $\sigma_r^2 = (Q^2 - Q_r^2)/\{(n-2)m\}$ (i.e., Equation 6.8) for $r = 1, 2$, and 3, where $m = 2$ is equal to the number of regions minus one, and where $n = 4$ is the number of time periods. We find $\sigma_r^2 = 2.51, 0.645$, and 0.673 for the three respective time periods.

Next, we calculate $K_r^2 = Q_r^2 / \sigma_r^2$ and find $1.376/2.51 = 0.542$, $8.83/0.645 = 13.69$, and $8.72/.673 = 12.96$, for $r = 1, 2$, and 3, respectively.

Finally, we compare the values of K_r^2/m with the critical value of an F distribution with m and $(n - 2)m$ degrees of freedom in the numerator and denominator, respectively. The values of K_r^2/m are $0.542/2 = 0.271$, $13.69/2 = 6.845$, and $12.96/2 = 6.48$, for $r = 1, 2$, and 3, respectively. The critical value of F with 2 and 4 degrees of freedom is 6.94; as none of our observed values exceeds this critical value, we fail to reject the null hypothesis, though just barely. The respective p-values associated with the three time periods are 0.7756, 0.0511, and 0.0556. Note also that in practice, σ_r^2 and K_r^2 need only be calculated for $\hat{r}$, because $Q_{\hat{r}}^2$ will always maximize $K_{\hat{r}}^2$.

6.4.2 Example 1: Breast Cancer Mortality in the Northeastern United States

The following example is adapted from Han and Rogerson (2003).

Annual data on observed breast cancer mortality frequency for the $m + 1 = 217$ counties of the northeastern United States were examined for the period 1968–1998.

Before computing the test statistic, it should first be noted that the spatial pattern of cases may sometimes shift due to changes in population, and it is desirable to adjust for such demographic shifts. A region may exhibit a dramatic increase in the number of cases, but this may be a direct result of a dramatic increase in population. Changes in age structure (and in other covariates) might also be taken into account. Such changes are irrelevant to the structure of disease occurrence, and thus, are not changes that we would be interested in.

To account for such "uninteresting" changes, the annual observed vectors of county deaths must be adjusted for changes in expectations, which are calculated to reflect changes in the population structure. This is done here by multiplying the observed number of county deaths in year t by the ratio of the proportion of all expected deaths (for the entire time period) that occur in county j among all the counties in the study area to the proportion of all expected deaths that occur in county j during year t. This adjustment discounts an observed number of deaths in a given county in a particular year if the county share in that year is more than the average share for that county.

The results are shown in Table 6.5. The changepoint statistic is significant, and indicates a change in the spatial pattern of breast cancer following 1989. The data are then divided into two subsets—1968–1989, and 1990–1998. Further tests on these two subsets indicate that the former period may be further subdivided into two distinct subperiods—1968–1976, and 1977–1989. The latter period (1990–1998) does not contain any significant changepoints. The 1977–1989 period may be further subdivided into 1977–1981, 1982–1986, 1987, and 1988–1989.

TABLE 6.5

Retrospective Detection of Changes in Breast Cancer
Mortality in the Northeastern United States, 1968–1998

Sequence	Change Point	K_r^2	F	p
1968–1998	1989	482.4	2.089	<.001
1968–1989	1976	354.0	1.64	<.001
1990–1998	1997	252.2	1.168	.059
1968–1976	1973	249.3	1.154	.075
1977–1989	1980	311.0	1.44	<.001
1977–1981	1976	239.3	1.11	.171
1982–1989	1986	286.7	1.33	.002
1982–1986	1985	217.9	1.01	.46
1987–1989	1987	298.3	1.38	.009

Homogeneous subsets: 1968–1976, 1977–1981, 1982–1986, 1987,
1988–1989, 1990–1998.

It is of interest to ask which counties have contributed most significantly
to the changes that have occurred. Table 6.6 depicts the number of deaths
occurring in selected counties for the 1968–1989 and 1990–1998 periods.
The final column of the table gives the ratio of the county's fractional share

TABLE 6.6

Ratio of County's Share of Deaths 1990–1998 to 1968–1989: Counties with Ratios
Greater than 1.2 or Less than 0.8

County	Deaths 1968–1989	Deaths 1990–1998	Ratio of County's Share of Deaths 1990–1998 to 1968–1989
Bennington, VT	174	56	0.746
Waldo, ME	115	36	0.725
Columbia, NY	282	96	0.787
Cortland, NY	177	52	0.680
St. Lawrence, NY	411	134	0.758
Wyoming, NY	157	51	0.746
Cameron, PA	33	10	0.676
Wyoming, PA	86	29	0.781
Oxford, ME	154	87	1.309
Ulster, NY	636	345	1.258
Fulton, PA	30	17	1.330
Juniata, PA	64	43	1.577
Potter, PA	56	43	1.770
Sullivan, PA	21	17	1.894
Sullivan, NH	129	81	1.466
Venango, PA	233	126	1.256
Warren, PA	153	87	1.271
Essex, VT	18	12	1.597
Orange, VT	69	43	1.451

of all deaths during 1990–1998 to its fractional share of all deaths during 1968–1989. Table 6.6 shows those counties with ratios greater than 1.2 or less than 0.8. Counties with ratios less than 0.8 experienced a substantial decline in breast cancer deaths during 1968–1989 to 1990–1998, relative to the rest of the study region. Those counties with ratios greater than 1.2, implying a substantial increase on the other hand, are generally small counties, with the exception of Ulster, New York.

6.4.3   Example 2: Recent Changes in the Spatial Pattern of Prostate Cancer Mortality in the United States

This example is adapted from Rogerson, Sinha, and Han (2006).

Rates of prostate cancer incidence and mortality have changed markedly over the last several decades. Incidence rates increased rapidly beginning in the late 1980s, following the widespread adoption of PSA (Prostate-Specific Antigen) screening. Incidence rates then began to fall in 1992. Prostate cancer mortality rates rose beginning in the late 1970s and through the 1980s, and fell slightly during the 1990s. Some work has attempted to establish a connection between improved screening and reduced mortality, but results to date have been mixed. The objective of this illustration is to examine the spatial pattern of changes that have taken place in prostate cancer mortality. Different regions have contributed differentially to the temporal changes in mortality just described, and this in turn may at least in part be due to regional differences in screening programs.

6.4.3.1   Introduction

Prostate cancer is the number one incident cancer and the second most common cause (after lung cancer) of cancer death for men in the United States. The American Cancer Society estimated that, in 2003, 220,900 men were diagnosed with prostate cancer and that 28,900 deaths were directly attributable to the disease (Crawford 2003). Epidemiologists have identified age, ethnicity, family history, and dietary practices as important risk factors.

Screening efforts became much more aggressive in the United States (and in some other Western countries) with the beginning of what is now commonly referred to as the PSA era (starting in 1986). The new screening methodology led to a rapid increase in incidence rates between 1986 and 1992; evidence for its effect on mortality rates is also becoming clearer. A survey of the current literature indicates that, at least in the United States, mortality rates have been decreasing steadily during the last decade. Recent research lends support to the theory that this decrease is due at least in part to PSA testing. However, caution is advised by some in interpreting these as long-term effects because the etiology of the disease is still largely unknown, and the effect of PSA testing on public health has yet to be shown conclusively.

In this illustration, we focus on the spatial changes in prostate cancer mortality that occurred during the period 1968–1998. Because of the relatively small number of nonwhites in many regions of the country, we confine our illustration to white males. The analysis is carried out at the county level and is based on the annual death counts attributable to prostate cancer.

6.4.3.2   Geographic Variation in Incidence and Mortality Rates

A small number of studies that focus upon the spatial pattern of prostate cancer find interesting and consistent patterns. The Atlas of Cancer Mortality (Devesa et al. 1999) revealed that prostate cancer mortality rates were higher among white men in the Northwest, Rocky Mountain States, New England, North-Central and South Atlantic areas, and for black males in the South Atlantic region. An inverse rural–urban gradient was suggested, with high rates in less populated areas. Similar patterns have been described by others. It has also been noted that recent patterns for white males are more clustered in the northwest than in earlier years. A recent study detected five statistically significant clusters for whites and three significant clusters for blacks in the United States; the patterns observed could not be explained away by the selected demographic and socioeconomic factors, indicating that further spatial analysis of the risk factors and medical or reporting practices is required (Jemal et al. 2002).

One explanation is that some of the geographic variability could be due to a possible protective effect of ultraviolet radiation. Bodiwala et al. (2003) point to a number of studies that suggest such a relationship, and they give additional evidence in support of a connection. The strong north–south gradient of prostate cancer rates in the United States, with high rates in the north and low rates in the south, is consistent with this hypothesis.

6.4.3.3   Data

Prostate cancer mortality based on death certificates was obtained from the Compressed Mortality File (CMF) produced by the National Center for Health Statistics. The CMF data are available at the county level for individual years for the period 1968–1998, grouped by age, sex, race, and underlying causes of death. The number of prostate cancer deaths (ICD-9 codes, 185.0–185.9) was obtained for each county in the contiguous United States from 1968 through 1998, for white men aged 25 and over, by 10-year age groups. We excluded 30 deaths with unknown age information.

We derived the expected number of prostate cancer deaths for white males using the population estimates that were based on Bureau of the

Census estimates of midyear county population and provided with the CMF. The expected number of prostate cancer deaths in each county was obtained using the indirect standardization method, by multiplying national age-specific death rates and the county population in each age group. The observed and expected numbers of prostate cancer deaths for each county in the United States were used as input for further spatial analyses as described in the next section.

6.4.3.4 Descriptive Measures of Change

The number of deaths due to prostate cancer among white males rose steadily and doubled during the period 1968–1994, with approximately 14,000 annual deaths at the beginning of this period, and slightly more than 28,000 by the end of the period. By 1998, the annual number of deaths had fallen slightly, to about 26,500.

We initially take a descriptive approach to evaluate temporal changes in the departure of mortality rates from spatial uniformity. For each year, multinomial probabilities (p_j) may be defined as the likelihoods that a given prostate cancer death is located in county j. A χ^2 statistic was computed for each year as a global measure of the difference between observed and expected county frequencies. To account for the fact that the number of deaths from prostate cancer essentially doubled during the time period, it is desirable to examine changes in the following quantity:

$$w = \sqrt{\sum_{j=1}^{m} \frac{(p_{obs} - p_{exp})^2}{p_{exp}}} = \sqrt{\chi^2/n} \qquad (6.9)$$

where m is the number of counties, n is the number of deaths from prostate cancer, and where p represents the observed and expected proportions of all national prostate cancer cases that fall in county j. The quantity w represents a measure of the global distance between the observed and expected multinomial probabilities. As an aside, when the p's in Equation 6.9 are replaced by the p's expected under null and alternative hypotheses, w becomes a measure of effect size (i.e., a measure of the distance between null and alternative hypotheses).

Figure 6.2 displays the annual values of w for the period 1968–1998 (in the figure, the vertical scale does not begin at zero, to exaggerate the vertical scale and allow for better visualization of temporal changes). Although there are no major changes over time, general trends are apparent. The figure reveals a fairly steady decline from the beginning of the study period until about 1993, indicating an increasing tendency toward spatial uniformity. The period 1994–1998 was one of increasing spatial concentration, and by 1998 the level of spatial concentration had returned to levels that prevailed during the late 1980s.

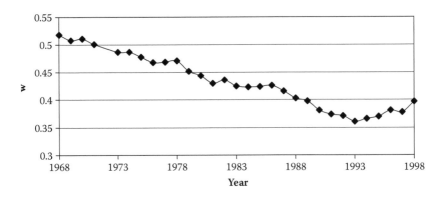

FIGURE 6.2
Distance from expectations (w): prostate cancer mortality in the United States, 1968–1998.

6.4.3.5 Retrospective Detection of Change

Before computing the test statistic, the annual observed vectors of county deaths were adjusted for changes in expectations that occurred as a result of population and age structure changes. As in the previous illustration, this was done by multiplying the observed number of county deaths in year t by the ratio of the proportion of all deaths (for the entire time period) that occur in county j to the proportion of all deaths that occur in county j during year t.

The results—in the form of the chi-square statistic associated with a breakpoint after each potential year—are shown in Figure 6.3. The change-point statistic, which is based upon the maximum chi-square value in the figure, is maximized and significant in 1990, indicating a change in the

FIGURE 6.3
Chi-square values for potential breakpoints.

spatial pattern of prostate cancer following that year. In addition, the maximum illustrated in the figure is quite peaked, instilling confidence in the accurate identification of the changepoint and indicating possible changes in the spatial pattern in a relatively narrow time window around 1990.

It is of interest to ask which counties contributed most significantly to the spatial changes that occurred. To assess this, for each county we first aggregated the annual observed and expected frequencies for each of the two time periods before and after the change (1968–1990 and 1991–1998). We next computed standardized z-scores associated with the aggregated observed frequencies for each county for each of these two periods, using the observed and expected frequencies:

$$z = \frac{f_{obs} - f_{exp}}{\sqrt{f_{exp}}} \qquad (6.10)$$

These z-scores have a mean of zero and a variance of one, and they represent a measure of how far the observed mortality frequencies are from expectations. They are based on a Poisson model for the distribution of observed deaths; f_{exp} represents both the mean and the variance of the Poisson distribution. The z-scores are constructed in the usual fashion by subtracting the expected value from the observed value and then dividing the result by the standard deviation. In addition, and more importantly from a conceptual perspective, the z-scores are equal in absolute value to the square root of the individual county's contribution to the chi-square statistic associated with the $2 \times m$ table of observed deaths (adjusted for changing expectations) that occurred in the m counties before and after the change.

For a given county, the difference in the two z-scores represents how the county's prostate cancer mortality (relative to expectations) has changed. This difference in z-scores has a variance of two (under the assumption that the covariance of the z-scores is equal to zero); hence the quantity

$$z = \frac{z_{91-98} - z_{68-90}}{\sqrt{2}} \qquad (6.11)$$

has, approximately, a standard normal distribution under the null hypothesis of spatial uniformity in both periods. (Although the assumption of normality is admittedly questionable for counties with small expected frequencies, the aggregation into two reasonably long time periods alleviates much of this concern.) Because the actual covariance of the z-scores may be positive (e.g., because if a county has mortality exceeding expectations before the change, it may also be likely to have mortality exceeding expectations after the change), the actual variance of the difference in z-scores may be greater than two. This in turn implies that the significance of these new z scores may be somewhat less than indicated here. However, we are primarily interested in the spatial distribution of the z values, and hence this limitation is not particularly troublesome.

Before mapping the z values, we first smoothed them to highlight spatial patterns; we used a Gaussian kernel, which corresponds to placing a normal distribution at the center of each county and then replacing the county's observation with the weighted sum of the observations in that county and surrounding counties (where the weights associated with each county are equal to the height of the normal distribution at that county's centroids; see Chapter 5, Section 5.10). The standard deviation of the normal distribution is associated with the amount of smoothing; larger values give substantial weight to even distant counties and result in a relatively large amount of smoothing. We chose as the standard deviation a distance equal to the square root of the average area of a county (corresponding roughly to the average length of a county's border). This choice corresponds approximately to the common practice of using adjacent counties in measures of spatial autocorrelation.

Figure 6.4 shows these standardized differences between the two time periods. The map depicts high positive values in western Appalachia and down through the south central portion of the country. These are areas where the observed number of prostate cancer cases has grown the most rapidly, with respect to the expected number of cases. Areas that display the most negative values include much of New England and scattered

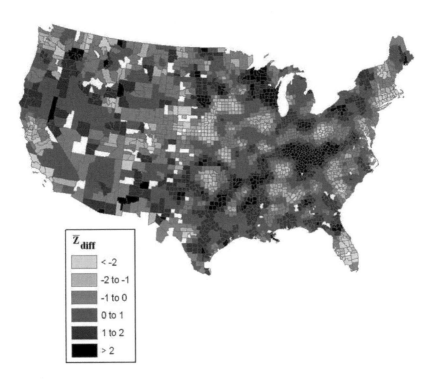

FIGURE 6.4
Standardized difference in z-scores between 1968–1990 and 1991–1998.

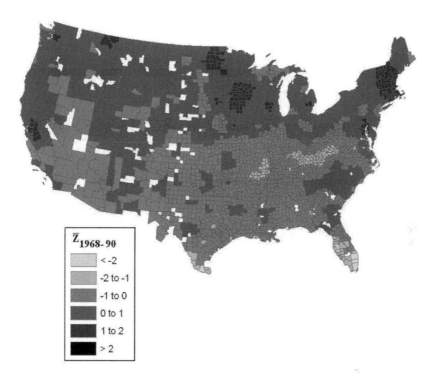

FIGURE 6.5
z-Score for 1968–1990.

portions of the upper Midwest and West; these are areas where the largest reductions in cancer mortality have occurred.

Figures 6.5 and 6.6, respectively, display the smoothed values of z_{68-90} and z_{91-98} (the difference of these maps is shown in Figure 6.4). It is interesting to note that many of the declines in mortality (shown as light areas in Figure 6.4) have occurred in the northern part of the country, where rates are highest. Likewise, many of the increases in mortality, relative to the expectations brought about by assuming spatial uniformity, have occurred in the southern portion of the country, where rates have historically been lower. In sum, the map of mortality observed for the aggregate period 1991–1998 shows less spatial variability than the map observed for the period 1968–1990. These results are consistent with the descriptive results presented earlier, which showed, from 1968 onwards, a steady march toward lesser geographic diversity and greater spatial uniformity, at least until about 1994.

The data for the period 1991–1998 were then examined using the same method, and a significant changepoint was discovered in 1994. The spatial differences in prostate cancer mortality between the 1991–1994 and 1995–1998 periods are highlighted in Figure 6.7. Although distinct spatial patterns are not as readily apparent, the declines in southern California and

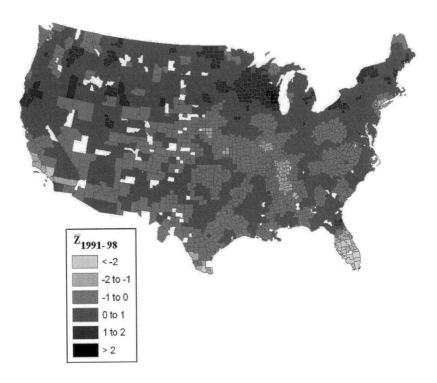

FIGURE 6.6
z-Score for 1991–1998.

Florida are notable because these constitute large urban areas; declines in these areas of already low prostate cancer mortality contributed to the decreasing spatial uniformity that occurred during the period 1995–1998.

It is also instructive to look at the locations of the maximum and minimum (unsmoothed) z-scores in each year; despite the limitations associated with looking at outliers, some trends emerge when examining these values over a 31-year period. The maximum standardized scores that were observed in each year are shown in Table 6.7. Also shown are the observed and expected counts for those counties where the maximum was observed. The scores in later years appear to be generally higher than those in earlier years.

The minimum standardized scores are shown in Table 6.8. It is interesting to note that the minima typically occur in large-population counties, whereas the maxima tend to occur in small-population counties; this is in keeping with a reverse rural–urban gradient that has been noted previously (with high values in rural areas and low value in urban regions). Also notable is the fact that Los Angeles County had the lowest score in 3 of the last 4 years for which data were available. It seems plausible that

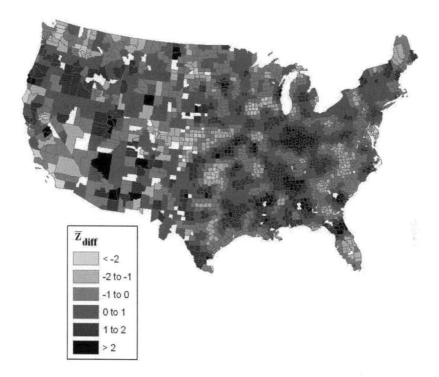

FIGURE 6.7
Standardized difference in z-scores between 1991–1994 and 1995–1998.

a part of the increasing spatial diversity in prostate cancer morality witnessed during the mid- to late-1990s may be due to the declining mortality in Los Angeles County, as well as declining mortality in parts of Florida. This would be consistent with what one might expect from the early implementation of screening programs; they are more likely to be adopted early in urban areas, and hence, any early improvements in mortality might be more likely to occur in such places.

6.4.3.6 Discussion

These spatial changes have occurred at about the time we would have expected changes to show up in mortality patterns, given an approximate time lag of 5 years from the start of widespread screening. Although there are alternative ways to interpret our results, one possible summary is

i. Spatial variation in prostate cancer mortality declined for a long period of time so that the pattern of deaths observed during the 1990s was considerably more uniform than that observed earlier. This was largely the result of long steady declines in spatial concentration that occurred until about 1994.

TABLE 6.7

Maximum z-Scores: Prostate Cancer Mortality in the United States, 1968–1998

Year	State	County	Observed	Expected	z
1968	ND	Dunn	6	1.2	3.55
1969	MA	Barnstable	17	10.3	3.52
1970	OR	Klamath	12	3.7	3.7
1971	NC	Davidson	15	5.54	3.53
1973	TX	Grimes	16	5.54	3.87
1974	MO	Wright	10	2.32	4.16
1975	ND	Barnes	8	1.97	3.58
1976	IA	Pocohantas	8	1.95	3.6
1977	MO	Bollinger	7	1.48	3.71
1978	MS	Scott	11	1.71	5.57
1979	MD	Baltimore City	80	49.28	4.11
1980	VT	Lamoille	3	1.39	3.78
1981	VA	Pittsylvania	13	3.96	3.89
1982	NE	Deuel	12	3.71	3.69
1983	IA	Calhoun	9	2.46	3.51
1984	ND	Pembina	9	1.72	4.47
1985	ID	Jefferson	8	1.44	4.37
1986	IA	Emmet	10	1.89	4.74
1987	IA	Webster	18	6.17	4.14
1988	NC	Oswego	7	1.12	4.36
1989	UT	Salt Lake	85	55.76	3.71
1990	MS	Choctaw	7	1.02	4.6
1991	MN	Benton	14	4.06	4.2
1992	IA	Bremer	14	4.1	4.16
1993	GA	Emanuel	9	2.03	4.03
1994	OR	Morrow	7	1.22	4.15
1995	MS	Madison	12	3.09	4.24
1996	WI	Kewaunee	13	3.18	4.57
1997	TN	Marion	12	2.77	4.57
1998	WI	Calumet	14	3.64	4.55

ii. Spatial concentration began increasing in 1994. An interesting and open question is whether these changes since about 1993 or 1994 can be attributed to spatial variability in the effectiveness of screening programs. This could come about, for instance, if screening programs were initially more widely adopted in areas of low mortality—and there is a possibility of this given that (a) there is some support for a reverse rural–urban gradient in prostate cancer mortality, and (b) it seems likely that screening would at least initially be more effective in urban than in rural areas. Spatial concentration would increase as a result of possibly stronger declines in mortality in urban areas (due to the early effects of screening programs), where mortality is already low relative to rural areas.

TABLE 6.8

Minimum Observed z-Scores: 1968–1998

Year	State	County	Observed	Expected	z
1968	WI	Douglas	0	5	−3.37
1969	FL	Pinellas	72	120.2	−4.68
1970	NY	Queens	123	171.92	−3.89
1971	NY	Bronx	79	115.08	−3.52
1973	NY	Kings	150	199.73	−3.64
1974	NY	Bronx	69	109.56	−4.1
1975	MO	Franklin	0	6	−3.67
1976	TX	Hidalgo	2	18.12	−4.74
1977	FL	Dade	140	189.31	−3.72
1978	VA	Augusta	0	7.62	−4.14
1979	PA	Warren	0	5.23	−3.43
1980	TX	Bell	1	9.67	−3.5
1981	NY	Kings	140	188.88	−3.69
1982	WI	Marinette	0	5.91	−3.65
1983	FL	Dade	166	215.88	−3.51
1984	FL	Broward	146	202.66	−4.14
1985	NY	Kings	141	190.25	−3.7
1986	NY	Queens	144	206.07	−4.52
1987	TX	Hidalgo	8	32.62	−5.04
1988	IL	Ogle	0	5.97	−3.67
1989	NY	Kings	149	200.49	−3.77
1990	TX	Hidalgo	16	40.67	−4.31
1991	NY	Kings	153	207.6	−3.93
1992	NY	Kings	144	106.92	−4.57
1993	FL	Broward	227	286.39	−3.61
1994	TX	Taylor	2	13.83	−3.9
1995	CA	Los Angeles	599	708.89	−4.21
1996	FL	Palm Beach	180	250.08	−4.61
1997	CA	Los Angeles	572	676.42	−4.1
1998	CA	Los Angeles	527	655.68	−5.16

There are some notable limitations to this analysis. The retrospective test for change in spatial patterns is based upon the alternative hypothesis that there is a single point in time that divides the spatial patterns into "before the change" and "after the change." Although Figure 6.2 seems to indicate gradual changes in descriptive measures of spatial concentration, Figure 6.3 seems to imply a more well-defined changepoint.

These figures therefore need to be interpreted with caution; they are not necessarily inconsistent but rather are simply limited in their ability to reveal a fuller picture of spatial change. It is possible to have a substantial change in geographic pattern with little or no change in systemwide measures of spatial concentration (this would occur, for example, if the individual counties where concentration was high shifted from one location to another). An important next step is to relate the results described here to changes in prostate cancer *incidence*. In particular, it would be of

TABLE 6.9

Burglary Incidents by Quarter in 1998, Buffalo

Period	Number
January–March	1364
April–June	1540
July–September	1645
October–December	1564
Total	6113

TABLE 6.10

Srivastava and Worsley's Maximal Chi-Square Statistic for the Buffalo Burglary Data

	χ_r^2	σ_r^2	$K = Q_r^2 / \sigma_r^2$	K/m	*p*-Value
Quarter 1 vs. Quarters 2,3,4	167.2	1.16	143.9	1.62	.004
Quarters 1,2 vs. Quarters 3,4	126.6	1.39	91.1	1.02	.44
Quarters 1,2,3 vs. Quarter 4	101.3	1.53	66.1	0.74	.94
Quarters 2, 3 vs. Quarter 4	107.3	1.13	94.8	1.07	.38
Quarter 2 vs. Quarters 3,4	85.8	1.37	78.1	0.70	.95

interest to know whether there was correspondence in the map depicting change in mortality and the map depicting change in incidence.

6.4.4   Example 3: Crime

This last example is adapted from Rogerson (2004a) and uses data on burglary incidents in the city of Buffalo, New York, for 1998. Information is available on the number of burglaries by census tract, as is information on the coordinates of tract centroids. In addition, the data are disaggregated by quarter; a summary of the number of incidents in each quarter is given in Table 6.9. There are 90 census tracts in the city, and there is an average of approximately 15 burglaries per tract, per quarter.

Results are shown in Table 6.10. The most likely "changepoint" is between the first quarter and the remainder of the year (*p*-value = 0.004); no other breakpoint yields a greater distinction between "before" and "after" patterns. The geographic pattern of burglaries during the fourth quarter is very similar to the pattern during the first three quarters (*p* = 0.94).

Once a changepoint has been found, it is then of interest to see whether other changepoints exist within either the "before" or "after" segments of the data. The bottom part of Table 6.10 reveals that further disaggregation of Quarters 2, 3, and 4 does not uncover additional, significant change in the spatial pattern of burglaries.

7

Introduction to Statistical Process Control and Nonspatial Cumulative Sum Methods of Surveillance

7.1 Introduction

Statistical process control (e.g., see Montgomery 1996, and Hawkins and Olwell 1998) refers to the statistical approaches that have been developed for monitoring a process (such as an industrial production process) over time so as to quickly detect any changes in the process. This has been carried out primarily in conjunction with the objective of improving industrial quality control; by quickly detecting changes that may affect the quality of the end product, one can correct the process before manufacturing defective products. Statistical process control methods generally utilize control charts to monitor an underlying variable of the process of interest for which repeated observations are available over time.

The methods of statistical process control have a long history of application to problems in public health surveillance. Hill et al. (1968) and Weatherall and Haskey (1976) were among the first to propose and implement such systems, with their applications to the temporal surveillance of congenital malformations. Barbujani (1987) provides a review of these methods. Farrington and Beale (1998) and Sonesson and Bock (2003) provide more recent and more general reviews of statistical surveillance in public health. Farrington and Beale (1998) provide a detailed summary of the motivation for *prospective*, as opposed to *retrospective*, detection of disease outbreaks. They discuss problems associated with the goals of prospective detection, including those associated with reporting delays, false detection, and the influences of past outbreaks and systematic (e.g., seasonal) variability. Farrington and Beale also review a number of methods that could be of use in prospective detection. Babcock et al. (2005) provide a more recent example of an application in this area.

Prospective disease surveillance has gained increasing attention, particularly in light of recent concern for quick detection of bioterrorist events. Monitoring has the potential for the detection of such events, but the benefits of surveillance extend much more broadly to the quick detection of changes in public health. In the field of public health, many of the current systems of this type focus on data collected prior to diagnosis, such as data on sales of over-the-counter medications and data on syndromes and symptoms as recorded in hospital emergency departments (e.g., see Mandl et al. 2003; Mostashari and Hartman 2003).

Methods of statistical process control include, but are not limited to, Shewhart charts, cumulative sum (cusum) methods, and exponentially weighted moving-average methods. Shewhart charts are designed to detect large deviations from the mean of a process; single, outlying observations can trigger an alarm or signal that the process mean may have changed. Cusum methods maintain a running total of the deviations between observed and expected values; if this total exceeds a predetermined threshold, an alarm is sounded, again indicating a potential change in the underlying mean of the process. In this chapter, we give some attention to Shewhart charts but focus primarily on the use of cusum methods. In the next chapter, we will pay particular attention to issues that arise when there is a desire to carry out surveillance in a multiregional setting.

We first review and illustrate the fundamentals of Shewhart charts and cusum methods. The discussion is general and is initially oriented toward variables that come from normal distributions. Much public health data are not normally distributed—especially small counts collected at frequent intervals, and data associated with uncommon diseases. Later in the chapter, we cover both transformations of such data to normality and methods designed to handle directly data that are other than normally distributed. We also point out several developments associated with cusum methods that have not been widely used in a public health context, and that may ultimately prove to be of value.

7.2   Shewhart Charts

Shewhart charts plot individual observations or the means of groups of observations as they are observed over time. Limits or thresholds are placed on the charts, and an "out-of-control" signal is sent if an observation is found to be outside these limits. When observations come from a standard normal distribution, the establishment of a threshold of ±3 would imply that, while the process is in control, any observation would cause a signal with probability of .0027 (as this is the area in the tails of the standard normal distribution, to the right of $z = +3$, and to the left of $z = -3$). This in turn implies a false alarm (where significant change is declared when it in fact has not occurred), on average, once every $1/0.0027 = 370$ observations. If false alarms were more, or less, tolerable, the threshold could be adjusted accordingly. For instance, if one allowed a false alarm every 100 observations, the acceptable probability of false alarms associated with each observation is $1/100 = 0.01$. Again assuming that observations come from a standard normal distribution, this probability is associated with a threshold of ± 2.576. On the other hand, if only one false alarm in every 1000 observations is allowed, the threshold is ± 3.291, so that a more extreme observation is needed for an alarm.

Although Shewhart charts are good at detecting large changes in the mean of a variable, they fare less well (e.g., in comparison with cusum methods) in the quick detection of more subtle changes in the mean.

7.2.1   Illustration

Figure 7.1 depicts two simulated time series observed over 30 time periods. The first 15 observations of both Series 1 and Series 2 were derived from the standard normal distribution $N(0,1)$; the remaining 15 observations were derived from $N(1,1)$ and $N(0.2,1)$ distributions, respectively. That is, from the 16th observation, the process mean was increased by 1.0 in Series 1 and by 0.2 in Series 2. Series 1 sounded an alarm indicating an increased process mean at time 22, when its observation first exceeded the upper threshold value of +3. On the other hand, Series 2 never sounded an alarm because no observation reached the threshold value. This simple example illustrates the difficulty of detecting a relatively small change in mean with Shewhart charts. It can be explained by the fact that the probability of an observation exceeding +3 is only 0.0228 for $N(1,1)$, and only 0.0026 for $N(0.2,1)$; the latter is only slightly larger than the probability of an observation exceeding +3 for $N(0,1)$, which is 0.0013.

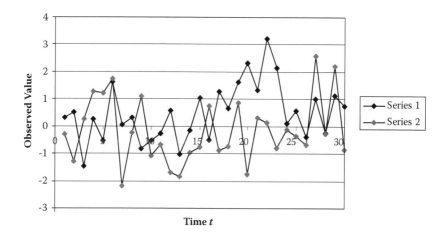

FIGURE 7.1
Shewhart charts for two time-series observations. Series 1: $N(0,1)$ for time 1–15; $N(1,1)$ for time 16–30; Series 2: $N(0,1)$ for time 1–15; $N(0.2,1)$ for time 16–30.

7.3   Cumulative Sum (Cusum) Methods

Cusum methods are designed to detect sudden changes in the mean value of a quantity of interest; they are widely used in industrial process control to monitor production quality (e.g., see Ryan 2000; Wetherill and Brown 1991; Montgomery 1996; Hawkins and Olwell 1998). The methods rely upon the assumption that the variable exhibits no serial autocorrelation. In the most common case, it is also assumed that the quantity being monitored is normally distributed, although it is possible and of interest to monitor observations that come from other distributions. We first review the case of normally distributed observations in this section. Some methods for observations distributed other than normal will be discussed subsequently.

The assumption of the normally distributed variable could be appropriate, for example, in the situation where we have monthly counts of the number of people with a particular disease in a large population. More specifically, if all individuals have the same probability of disease, the underlying distribution of disease counts is binomial, but this can often be approximated by the normal distribution for a large population. Without loss of generality, let the variable of interest (e.g., disease counts) be converted to a z-score with mean zero and variance one (by first subtracting the expected mean, and then dividing the result by the standard deviation). For binomial counts, $z_t = (O_t - np)/\sqrt{np(1-p)}$, where n is the size of the population, p is the probability of disease, and O_t is the observed number of cases at time t. For Poisson counts we use $z_t = (O_t - E_t)/\sqrt{E_t}$, where O_t and E_t represent observed and expected counts at time t, respectively. The expected count refers to the expectation under the null hypothesis that the process is "in control," that is, there is no increase in the disease occurrence relative to expectations. The cusum, following observation t, is defined as

$$S_t = \max(0, S_{t-1} + z_t - k) \qquad (7.1)$$

where k is a parameter sometimes called the *reference value*, and the cusum is started at zero (i.e., $S_0 = 0$). A change in mean is signaled if $S_t > h$, where h is a threshold parameter to be prespecified as described later. Signals will sometimes occur when no actual change has taken place; the expected time until a false alarm is called the "in-control" average run length, and it is designated by the notation ARL_0. Note that this cusum definition is for a one-sided test to detect increases in the mean; although a cusum method can be configured to detect decreases in the mean, we focus on this definition in this chapter because, in the context of public health studies, the detection of increased disease occurrence is of general interest.

Note that values of z in excess of k are cumulated in Equation 7.1. The parameter k in this instance, where a standardized variable is being monitored, is often chosen to be equal to ½; in the more general case, where the variable of interest may not have been standardized, k is often chosen to be equal to ½ the standard deviation associated with the variable being monitored. The choice of $k = ½$ minimizes the average out-of-control run length (that is, the average number of observations between the time of change and the time of detection) for a given value of ARL_0, when a true increase of one standard deviation has occurred. More generally, k is chosen to minimize the time needed to detect a change of $2k$ in the mean of the standardized variable.

The threshold parameter h is chosen in conjunction with a predetermined, acceptable rate of "false alarms"; high values of h will infrequently lead to false alarms, but it will also take a relatively long time to detect real change. Similarly, low values of h will yield a scheme that is quicker at detecting change when it occurs, but false alarms will be more frequent when no change has occurred. Table 7.1 depicts in-control average run lengths (ARL_0) for various values of h and $k = 0.5$, that is, the average times until a false alarm when the process is indeed in control. When a nonstandardized variable is being monitored, the critical value of the cusum is determined by multiplying the value of h by the standard deviation of the variable being monitored.

Most texts on statistical process control have tables and charts that may be used to find the value of h that is associated with chosen values of ARL_0 and k. An approximation for ARL_0 is provided by Siegmund (1985):

$$ARL_0 = \frac{\exp(2k(h+1.166)) - 2k(h+1.166) - 1}{2k^2} \tag{7.2}$$

In the special case when $k = ½$, this becomes

$$ARL_0 = 2(e^a - a - 1) \tag{7.3}$$

where $a = h + 1.166$. One can make practical use of this approximation to choose the parameter h by first deciding upon values of ARL_0 and k, and then the approximation may be solved numerically for the corresponding value of h. For example, for $h = 2.8$ and $k = ½$, Equation 7.2 returns 95.614, indicating that a false alarm is expected for every 95–96 observations. Because the distribution of average run lengths is approximately exponential, this implies that the choice of $h = 2.8$ corresponds to an approximate probability that a given observation leads to an false alarm of $1/95.614 \approx 0.01$. This approximation works well except in instances where k is large enough to yield h values in the neighborhood of zero.

Siegmund's approximation requires numerical methods to solve for the threshold parameter h, for a desired and specified value of ARL_0. Rogerson (2006b) has recently shown that Siegmund's equation may be solved, approximately, for the threshold parameter as a function of the

TABLE 7.1

In-Control ARLs (False-Alarm Rates Denoted ARL_0) for Various Values of h and $k = 0.5$ for Equation 7.1

h	ARL_0
2.5	68.9
2.6	76.9
2.7	85.8
2.8	95.6
2.9	106.5
3.0	118.6
3.1	131.9
3.2	146.7
3.3	163.1
3.4	181.2
3.5	201.2
3.6	223.4
3.7	247.9
3.8	275.0
3.9	304.9
4.0	338.1
4.1	374.7
4.2	415.3
4.3	460.1
4.4	509.6
4.5	564.4
4.6	625.0
4.7	691.9
4.8	766.0
4.9	847.8
5.0	938.2
5.1	1038.2
5.2	1148.7
5.3	1270.9
5.4	1405.9
5.5	1560.0

in-control average run length. For $k = \frac{1}{2}$,

$$h \approx \left(\frac{ARL_0 + 4}{ARL_0 + 2} \right) \ln \left(\frac{ARL_0}{2} + 1 \right) - 1.166 \qquad (7.4)$$

When k is not necessarily equal to $\frac{1}{2}$, the more general form of the equation for h is

$$h \approx \left(\frac{2k^2 ARL_0 + 2}{2k^2 ARL_0 + 1} \right) \frac{\ln(1 + 2k^2 ARL_0)}{2k} - 1.166 \qquad (7.5)$$

For instance, if we wanted to carry out surveillance corresponding with a probability of 0.005 that an individual observation caused an alarm, the in-control ARL_0 should be $1/0.005 = 200$. By using Equation 7.4, the appropriate h value is obtained as 3.49 for $k = \frac{1}{2}$.

7.3.1   Illustration

To illustrate how the cusum methodology may be implemented, consider the data in Table 7.2. The data were developed by first choosing random variates from a standard normal distribution for the first 15 time periods. Beginning in period 16, the mean increased by 0.2.

TABLE 7.2

Hypothetical z-Score Observations Following $N(0, 1)$ for Time 1–15 and $N(0.2, 1)$ for Time 16–30

Time Period	z-Score
1	−0.8
2	0.3
3	0.31
4	1.76
5	−0.75
6	0.38
7	0.57
8	−0.79
9	−1.04
10	1.14
11	1.4
12	0.35
13	1.38
14	−0.01
15	1.56
16	−0.57
17	1.98
18	1.77
19	−0.89
20	−0.81
21	−0.52
22	1.18
23	0.55
24	0.63
25	−0.2
26	−0.44
27	0.65
28	−1.99
29	1.28
30	0.06

TABLE 7.3

Cusum Computation for the Data in Table 7.2

Time t	$S_{t-1} + z_t - k$	$S_t = \max(0, S_{t-1} + z_t - k)$
1	−1.3	0
2	−0.2	0
3	−0.19	0
4	1.26	1.26
5	0.01	0.01
6	−0.11	0
7	0.07	0.07
8	−1.22	0
9	−1.54	0
10	0.64	0.64
11	1.54	1.54
12	1.39	1.39
13	2.27	2.27
14	1.76	1.76
15	2.82	2.82
16	1.75	1.75
17	3.23	*3.23*
18	4.5	*4.5*
19	3.11	*3.11*
20	1.8	1.8
21	0.78	0.78
22	1.46	1.46
23	1.51	1.51
24	1.64	1.64
25	0.94	0.94
26	0	0
27	0.15	0.15
28	−2.34	0
29	0.78	0.78
30	0.34	0.34

Note: Italic characters indicate cusum values greater than the threshold value (2.84).

If we use $k = 0.5$, and assume a desired ARL_0 of 100, Equation 7.4 implies a threshold of $h = 2.84$. Because average run lengths have distributions that are approximately exponential, note that this particular choice of ARL_0 corresponds to a .01 probability that an individual observation leads to an alarm.

Table 7.3 depicts how the cusum values are updated over time. The first observation is −0.8, so that $S_{t-1} + z_t - k$ is −1.3 (as a reminder, $S_0 = 0$). Because this is smaller than zero, the cusum value at time 1 is zero. The cusum value first exceeds the threshold value of 2.84 at time 17, for which $S_{t-1} = 1.75$ and $z_t = 1.98$, resulting in $S_{t-1} + z_t - k = 3.23$. Cusum values are often evaluated by means of a graph like Figure 7.2, which also includes a plot of the original observed values (i.e., Shewhart chart) for comparison

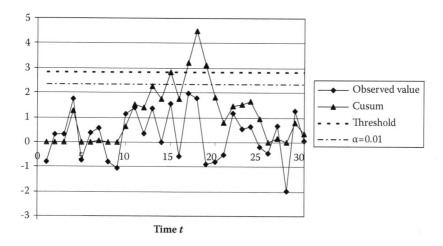

FIGURE 7.2
Observed values and the associated cusum chart for the data in Table 7.2.

purposes. Although the cusum value exceeds the threshold value at time 17 and also for the subsequent two time periods, the observation never exceeds the normal standard threshold for $\alpha = 0.01$ (i.e., 2.33), implying that the Shewhart chart cannot detect any change in this example. Furthermore, even if the observation at time 17 had been as low as 1.59, the cusum value would still exceed the threshold and signal a potential increase in the mean. It can also be pointed out that the observation at time 4, when the process is in fact in control, comes relatively close to the threshold value of Shewhart chart, whereas the cusum value at time 4 is less than half of the corresponding threshold. This comparison suggests the potential of the cusum method to detect mean changes more quickly than the Shewhart chart when the same significance level is used, as well as its robustness to random variation in in-control observations.

7.4 Monitoring Small Counts

Many situations call for the monitoring of data consisting of small counts. For example, data on the number of daily traffic accidents at an intersection, the number of cases of a rare disease, or the monthly number of violent crimes in a small city all generate sequences of small counts. Such data necessitate approaches to surveillance that are different from those discussed in the previous section because the assumption of normality is no longer tenable.

If the expected number of cases in a region is too small for the normality approximation to be tenable, there are several alternatives. One option is to "batch" observations for a small number of periods because the sums of binomial or Poisson variables observed over a number of periods will be closer to normally distributed than the original ones. An example of this strategy would be to monitor weekly counts instead of daily counts. This approach has the disadvantage that the timeliness of detection may be compromised by the need to accumulate enough observations for batching. Another alternative is to transform the binomial or Poisson variable to normality (Blom 1954). Finally, it is possible to use a binomial or Poisson cusum chart (Lucas 1985; Bourke 2001). Binomial and Poisson cusum charts are designed for the case where the underlying probability (p) or mean intensity (λ) is constant, and the purpose is to detect a sudden shift to values greater than p or λ. It is possible to modify them to include the more general case where the expected probability under the null hypothesis is shifting over time (Rogerson and Yamada 2004b), as is the case in the bioterrorism surveillance application described by Kleinman et al. (2004). More detail on these alternatives is given in subsequent subsections.

7.4.1   Transformations to Normality

When the assumption of normality for the distribution of observations is questionable, transformations of the observations to approximate normality may be possible. Instead of monitoring the original variable, the transformed variable is monitored. One such normalizing transformation for data consisting of small counts is (Rossi et al. 1999):

$$z = \frac{O - 3E + 2\sqrt{EO}}{2\sqrt{E}} \tag{7.6}$$

where O is the observed count and E is the expected count. The transformed values will have, approximately, a normal distribution, with a mean of zero and a variance of one.

As Rogerson and Yamada (2004b) indicate, this transformation can be misleading for very small values of E (say, of less than about two). In particular, the actual ARL_0 values may differ substantially from the desired nominal values. For example, when desired values of $ARL_0 = 500$ and $ARL_1 = 3$ (where ARL_1 is the average time taken to detect an increase) are used in situations where $E < 2$, simulations of in-control processes show that using this transformation will almost always yield actual values of ARL_0 significantly lower than the desired value of 500. In some cases (e.g., $E \approx 0.15$), the actual ARL will even be lower than 100, indicating a much higher rate of false alarms than desired. The performance is better when $ARL_0 = 500$ and $ARL_1 = 7$, but use of the transformation will again lead to substantially more false alarms than desired when E is less than about 0.25. Also troubling is the instability with respect to similar values of E; $E = 0.56$ will lead to an ARL_0 of around 400 when transformed observations are monitored, whereas $E = 0.62$ is associated with an ARL_0 of over 700. This is also true when $ARL_1 = 3$: $E = 0.96$ has an ARL_0 of about 212, whereas $E = 0.98$ has an ARL_0 of 635!

7.5   Cumulative Sums for Poisson Variables

Hawkins and Olwell (1998) point out that, even when the transformation to normality is reasonably accurate, the idea of monitoring transformed variables is suboptimal with respect to the alternative of using a properly designed cusum scheme for the untransformed data. Thus, although the normalizing transformation is convenient, there is reason to be cautious when applying it because the times to detection for Poisson data will be shortest when the proper Poisson cusum procedure is employed. Using the cusum procedure for normal variables on the transformed data will generally lead to longer (though not usually substantially longer) detection times.

In this section we review the Poisson cusum (Lucas 1985), which is appropriate for the monitoring of small counts in instances either where transformations are not adequate and/or where optimality (in terms of the quickest possible detection time) is very important.

There have been several applications of Poisson cusums in a public health context; examples include the surveillance of congenital malformations (Hill and Weatherall 1968; Weatherall and Haskey 1976), *salmonella* outbreaks (Hutwagner et al. 1997), and lower respiratory infection (Rogerson and Yamada 2004b).

7.5.1   Cusum Charts for Poisson Data

Surveillance of public health data often requires methods that are able to handle the monitoring of rare events effectively. When the variable being monitored has a Poisson distribution, the Poisson cusum (Lucas 1985) defined as follows can be utilized:

$$S_t = \max(0, S_{t-1} + y_t - k) \tag{7.7}$$

where y_t is the count observed at time t. We now discuss determination of the parameters k and h. Let $\lambda^{(a)}$ be the mean value of the in-control Poisson parameter. Following Lucas, the corresponding k-value that minimizes the time to detect a change from $\lambda^{(a)}$ to the specified out-of-control parameter ($\lambda^{(d)}$) is determined by

$$k = \frac{\lambda^{(d)} - \lambda^{(a)}}{\ln \lambda^{(d)} - \ln \lambda^{(a)}} \tag{7.8}$$

Then the threshold parameter h may be found from the values of the parameter k and the desired ARL_0 by either using a table (e.g., see Lucas 1985), Monte Carlo simulation, or an algorithm such as the one provided by White and Keats (1996), which makes use of a Markov chain approximation. To illustrate, if $\lambda^{(a)} = 4$, one might desire to detect quickly a one standard deviation increase to $\lambda^{(d)} = 6$; in this case, we would first find $k = 4.93$ from Equation 7.8. Then, if we desired an ARL_0 of approximately 420, we could use Table 2 as found in Lucas (1985) to find $h = 10$.

7.5.1.1   Example: Kidney Failure in Cats

In early 2007, it was reported that kidney failure among cats in the United States rose 30%. This increase was ultimately linked with pet food that had been contaminated with an industrial chemical. The affected food was packaged between December 3, 2006, and March 6, 2007, and was recalled on March 16. An average of three cats per day developed kidney failure during the 3-month period that the contaminated food was on the market. First complaints were received by the producer from pet owners on February 20, and testing began February 27.[*][+]

We now ask how quickly this increase might have been detected had the data been subject to ongoing monitoring. Suppose we had implemented a Poisson cusum with a daily expected count of $\lambda_0 = 2.3$, which is approximately 30% lower than the rate seen during the outbreak. Further suppose that we had fortuitously chosen $\lambda_1 = 3$ (i.e., the average daily number of cats that developed kidney failure during the outbreak), thereby minimizing the time to detection. We ignore here daily variability in the baseline number of visits (e.g., ignoring the fact that visits are much lower on weekends). Equation 7.8 implies that $k = 2.635$. Assume that we wish about 5 years between false alarms. Monte Carlo simulation reveals that, for the in-control process, a choice of $h = 17$ yields an ARL_0 of about 1600 days. Simulation also shows that when the daily rate increases to 3.0, the mean time to detection (ARL_1) is about 41 days. Note that this detection time represents a substantial improvement over the actual time period of 79 days—from December 3, 2006, to February 20, 2007.

---

[*] http://www.cbsnews.com/stories/2007/03/20/national/main2587087.shtml?source= mostpop_story.
[+] http://www.fda.gov/bbs/topics/NEWS/2007/NEW01590.html.

If more false alarms were tolerable, outbreak detection could have occurred more quickly. For example, a threshold of $h = 15$ yields a false-alarm rate of about $ARL_0 = 900$, and a mean time to outbreak detection of about $ARL_1 = 36$ days.

It is unlikely that any monitoring would have been "tuned" correctly (i.e., that the choice of k would have been optimal). Suppose that a larger shift in daily rate had been expected, say to $\lambda_1 = 3.8$, which corresponds approximately to a 65% increase from the baseline count λ_0 of 2.3. This yields $k = 2.988$. A threshold of $h = 11$ yields a false-alarm rate of about $ARL_0 = 1650$ (because of the discreteness of the Poisson distribution, it is not possible to achieve specific ARLs exactly). These parameters imply an average detection time of about 53 days—not as good as 41 days, but still significantly better than 79. Using $\lambda_1 = 2.7$ leads to $k = 2.50$, and $h = 22$ yields the desired ARL_0 (approximately 1500 in this particular case). These parameters lead to an average detection time of 43 days, only slightly worse than the 41 days found earlier.

7.5.2 Poisson Cusums with Time-Varying Expectations

The expected, in-control value associated with the Poisson variable may vary with time $(\lambda_{0,t}; t = 1, 2, ...)$, due, for example, to seasonal effects or weekday-versus-weekend variations in hospital visits. If the actual values of λ_0 fluctuated from period to period about the constant assumed parameter, simply applying a cusum scheme with constant parameters would yield misleading results.

A way of taking into account temporal variations in in-control parameters is to use values of the parameters k and h that are time specific. The observed values y_t may then be used in a modified cusum scheme defined as follows:

$$S_t = \max(0, S_{t-1} + c_t(y_t - k_t)) \tag{7.9}$$

where the parameters c_t and k_t change from one period to the next. Determination of the values of these two parameters is explained below.

First, a systemwide (i.e., not time-specific) value of k is determined based upon the mean of the time-varying in-control Poisson parameter $\lambda_{0,t}$ and a desired ARL_0; a systemwide threshold h is then chosen accordingly. Once h is chosen, k_t for a specific time t is next chosen based upon the in-control and out-of-control Poisson parameters, denoted by $\lambda_{0,t}$ and $\lambda_{1,t}$, respectively:

$$k_t = \frac{\lambda_{1,t} - \lambda_{0,t}}{\ln \lambda_{1,t} - \ln \lambda_{0,t}} \tag{7.10}$$

The next step is to determine the time-specific value of the threshold h_t that is associated with the desired ARL_0, k_t, $\lambda_{0,t}$, and $\lambda_{1,t}$. Finally, c_t is chosen as the ratio h to h_t, that is, $c_t = h/h_t$. The quantity c_t is a scaling factor that is chosen so that observed counts y_t will make the proper relative contribution toward the systemwide signaling threshold h used in the actual cusum. If, for example, the in-control and out-of-control Poisson parameter of a particular time period t results in a threshold value less than the one for the average Poisson parameter (i.e., $h > h_t$), then the contribution $y_t - k_t$ is scaled up by the factor h/h_t.

For instance, let us consider a situation where the average Poisson parameter over time λ_0 is 2.0, and the time-specific parameter for time t is $\lambda_{0,t} = 2.5$. Further assume that we would like to detect an increase of one standard deviation as quickly as possible with ARL_0 equal to 200. For the average Poisson parameter λ_0, the associated out-of-control parameter λ_1 is 3.41, and the value of k is derived as 2.64 by Equation 7.8. Using the Markov-chain-based algorithm provided by White and Keats (1996), an appropriate h value to achieve $ARL_0 = 200$ is determined to be 6.375. The out-of-control parameter $\lambda_{1,t}$ associated with $\lambda_{0,t} = 2.5$ is 4.08 (which is derived from $\lambda_{0,t} + \sqrt{\lambda_{0,t}} = 2.5 + \sqrt{2.5}$), and Equation 7.10 returns $k_t = 3.23$; the same algorithm provides $h_t = 6.889$. Here, $h < h_t$, implying that if the h value was used for an observed count at time t, false alarms would be sent more frequently than desired. Therefore, when the cusum scheme defined by Equation 7.9 is applied, the observation at time t needs to be discounted by the factor of $c_t = h/h_t = 6.375/6.889 = 0.925$.

An alternative approach suggested by Hawkins and Olwell (1998) is to apply a multiplicative factor to the "baseline" or average value of λ.

7.5.2.1   Example: Lower Respiratory Infection Episodes*

Harvard Vanguard Medical Associates (Boston, Massachusetts) uses an automated record system for its 14 clinics in the Boston area, and this case study utilizes data on lower respiratory infection episodes for January 1996–October 1999 extracted from the automated system. This study period contains 47,731 episodes that could be geocoded to one of the 287 census tracts composing the study region on the basis of patients' address records. Data on the number of visits to clinics made from each of 287 census tracts are monitored for the first 303 days of 1999, based upon expectations formed using daily data for the period 1996–1998. Because the expectations of daily counts, which were estimated as a function of month, a dummy weekend/weekday variable, and a time trend, were not constant, the Poisson cusum parameters, k and λ of each tract should also vary over time. Therefore, to form the Poisson cusum, Equations 7.9 and 7.10 were used to reflect the temporally varying nature of the data.

---

* From Rogerson and Yamada 2004b.

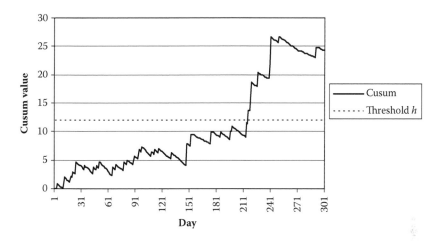

FIGURE 7.3
Poisson cusum chart for a selected census tract.

Figure 7.3 shows the Poisson cusum for one of the census tracts. During the base period, this tract had an average of 0.12 cases per day; this rose to 0.135 cases/day during 1999. The cusum crosses the threshold of $h = 12$ in early August of 1999, around the 220th day of monitoring. Cases leading to the alarm occurred on August 4th, 6th, and 9th (there were 2 cases observed on the 9th). These four cases in 6 days (0.67 cases/day) caused the Poisson cusum to rise above the critical threshold.

7.6   Cusum Methods for Exponential Data

An alternative approach for monitoring rare events is to use the fact that the times between Poisson-distributed events are exponentially distributed. Chen and her colleagues have a series of papers on the Sets method, which assumes exponential waiting times for events in a homogeneous Poisson process (Chen 1978; Chen et al. 1993; Chen et al. 1997). Sonesson and Bock (2003) discuss extensions to heterogeneous Poisson processes.

Lucas (1985) discusses the exponential (or time-between-events) cusum, and Gan (1994) compares its performance with the Poisson cusum. Gan finds that when there is interest in detecting an increase in the frequency of events, the exponential cusum outperforms the Poisson cusum, especially when there are large changes in the event frequency. This is because the Poisson cusum does not signal until the end of the unit time period; the exponential cusum is able to capitalize on the data it uses by signaling *during* a period in which the frequency has increased. For example, to calculate a daily Poisson cusum value, one needs to wait until the end of the day, whereas a time-between-event cusum can be updated every time an event occurs during the day. Similarly, Wolter (1987) notes that monitoring the gaps between events is more efficient than monitoring the number of events per time period when the latter is small. Borror et al. (2003) have recently shown that the exponential cusum is relatively robust with respect to departures from the assumption that the underlying distribution is exponential.

For an exponential distribution with mean $1/\theta$, that is,

$$f(x) = \theta \exp(-\theta x) \tag{7.11}$$

a potential change from an in-control value of θ_0 to an out-of-control value θ_1 can be monitored by first defining

$$k = \frac{\theta_1 - \theta_0}{\ln(\theta_0 \theta_1)} \tag{7.12}$$

The following cusum scheme is designed to detect an increase from θ_0 to θ_1 where $\theta_1 > \theta_0$, which corresponds to a decrease in the mean time between events and thus an increase in the frequency of events:

$$S_t = \max(0, S_{t-1} + 1 - kx_t) \tag{7.13}$$

where x_t is the time between events $t-1$ and t. An alternative expression for the cusum is

$$S_t = \sum_{i=1}^{t}(1 - kx_i) = t - k\sum_{i=1}^{t} x_i \tag{7.14}$$

To determine the threshold associated with the calculated value of k and a desired ARL_0, Gan (1994) provides charts (or nomographs). An alternative approach suggested by Alwan (2000) is to transform the data to normality by raising the observed x values to the power 0.2777 (i.e., $y = x^{0.2777}$). Alwan also gives the expectation and variance associated with these transformed values:

$$E[y] = .9011\theta_0^{-0.2777}$$

$$V[y] = .2780\theta_0^{-0.2777} \tag{7.15}$$

This allows one to implement the more common cusum based on the assumption of normality (along with approximations such as Equation 7.4 for determining the appropriate threshold).

Hawkins and Olwell (1998) emphasize that transformations will adversely affect the performance of the cusum. In this case, the normality transformation will increase the time to detection when a change has occurred (although Alwan shows that the effect is not large). It is therefore of interest to determine the threshold for the exponential cusum directly. Without loss of generality, the problem is transformed into one having an in-control parameter of 1 and an out-of control parameter equal to $\theta_1 = \theta_1/\theta_0$. This is achieved by normalizing the observed values by dividing each by θ_0. Based on the work of Siegmund (1985), it is possible to derive

$$\text{ARL}_0 \approx \frac{e^{\ln(\theta_1)(h+1.33)} - \ln(\theta_1)(h+1.33) - 1}{\ln(\theta_1)\,|1-k|} \tag{7.16}$$

Using arguments similar to those in Rogerson (2006b), this equation may be solved approximately for the threshold h in terms of ARL_0:

$$h \approx \frac{q+2}{q+1}\frac{\ln(q+1)}{\ln(\theta_1)} - 1.33 \tag{7.17}$$

where

$$q = \text{ARL}_0\ln(\theta_1)\,|1-k| \tag{7.18}$$

when one wishes to monitor for a decrease in the mean of the exponential variable (i.e., an increase in the parameter θ), which implies an increase in disease occurrence. Instead, monitoring for an increase in the mean of the exponential variable (i.e., a *decrease* in the parameter θ), the threshold can be approximated by

$$h \approx \text{ARL}_0(k-1) + \frac{1}{\ln(\theta_1)} - 1.33 \tag{7.19}$$

7.6.1 Illustration

Consider the data series shown in the second column of Table 7.4. The first ten observations were generated from an exponential distribution with parameter $\theta_0 = 1$ (mean equal to one); and the next ten were generated from an exponential distribution with parameter $\theta_1 = \theta_1 = 2$ (mean equal to one-half). Suppose we monitor for deviations from $\theta_0 = 1$ and, based on a desire to minimize the time to find a change to $\theta_0 = 2$, we use $k = (2 - 1)/$

TABLE 7.4

Exponentially Distributed Time Series and Associated Cusum Values

Observation	Time Since the Previous Observation (x_t)	Cusum Value (S_t)
1	0.284	1.159
2	2.608	0
3	0.289	1.154
4	0.730	1.867
5	1.255	2.055
6	0.205	3.293
7	4.162	0.574
8	0.077	1.940
9	0.245	3.138
10	0.981	3.600
11	1.110	3.933
12	0.299	5.077
13	0.086	6.434
14	0.389	7.488
15	0.046	8.885
16	0.224	10.104
17	0.220	11.327
18	0.666	12.104
19	0.002	13.545
20	0.275	14.713

$\ln(2) = 1.443$, based on Equation 7.12. If we use $\text{ARL}_0 = 100$, from Equation 7.18, we have $q = 30.69$, and then, from Equation 7.17, we find $h = 3.813$. Using Equation 7.13, one can compute cusum values (S_t) as shown in the third column of Table 7.4. The threshold value of $h = 3.813$ was first exceeded at time 11, sending a signal of an increase in the mean. Note that in Equation 7.13, an observed value x_t is subtracted from the cusum value at time $t - 1$. A very short time between two consecutive events might be an indicator of rate increase and should positively impact on the cusum value. This is why the observed time between two events is subtracted from the previous cusum value, differently from other cusum schemes discussed in previous sections where an observed value is added to the previous cusum value.

7.7 Other Useful Modifications for Cusum Charts

7.7.1 Fast Initial Response

One common modification is to start the cusum at a value other than zero. Lucas and Crosier (1982) recommend starting the cusum at a value of $h/2$ instead of zero. This has the benefit of signaling changes much more quickly if the observations begin in an out-of-control state. This benefit, called the fast initial response (FIR) feature, is achieved at the cost of either (a) a slightly higher rate of false alarms if the usual value of h is used, or (b) the value of h is raised slightly to maintain the desired ARL_0, in which case the time to detection is slightly longer than it would have been without using the FIR feature. The FIR cusum is employed widely because the benefit of quicker detection of change in out-of-control startup situations generally outweighs the small costs just described.

Let us consider the data series shown in Table 7.2. The appropriate h value for $ARL_0 = 100$ is 2.84, so that the FIR cusum starts with $S_0 = h/2 = 1.42$. For the first observation $z_1 = -0.8$, $S_0 + z_1 - k = 1.42 + (-0.8) - 0.5 = 0.12$; thus $S_1 = \max(0, S_0 + z_1 - k) = 0.12$, whereas $S_1 = 0$ without the FIR feature as shown in Table 7.3. For the second observation $z_2 = 0.3$, $S_1 + z_2 - k = 0.12 + 0.3 - 0.5 = -0.08$, and thus, $S_2 = 0$, which is the same value for the regular (i.e., without FIR) cusum. In this particular situation, the FIR feature does not improve the timeliness of detecting the change. However, if the surveillance process started at time 16, the cusum with FIR would have detected the change at time 18, whereas the cusum without FIR would have detected no changes. This illustration also suggests that, regardless of the use of FIR, results of the cusum surveillance could be impacted strongly according to when the surveillance had been started.

7.7.2 Unknown Process Parameters

Perhaps even more important in the context of public health surveillance is the fact that the "in-control" parameters are often not known, and are instead estimated from recent or historical data. For example, the "true" rate of malformations in a health authority's geographical area that is to be used as a baseline expectation for the future rate of malformations is most often based upon either recent data for that area or some larger geographic area.

Hawkins and Olwell (1998) demonstrate that, if an historical sample is used to estimate an unknown mean, subsequent surveillance can have false-alarm rates that are much higher or much lower than the desired, nominal value of ARL_0. This is because the "true" rate is not known. To account for this, they suggest a self-starting approach that may be summarized as follows (for normally distributed variables; other self-starting approaches also may be devised for variables with other distributions). As each observation X_n is made, the quantity

$$T_n = \frac{X_n - \bar{X}_{n-1}}{s_{n-1}} \tag{7.20}$$

is found, where $\bar{X}_{n-1}$ and s_{n-1} are the sample mean and standard deviation based upon the first $n - 1$ observations. The quantity

$$V_n = \sqrt{\frac{n-1}{n}} T_n \qquad (7.21)$$

has a t-distribution with $n - 2$ degrees of freedom. This can be transformed into a quantity that has a standard normal distribution as follows:

$$U_n = \Phi^{-1}[F_{n-2}(V_n)] \qquad (7.22)$$

where F_{n-2} is the cumulative distribution function for the t-distribution with $n - 2$ degrees of freedom, and Φ^{-1} is the inverse of the normal distribution. Thus, the U_n's represent the value from a standard normal distribution that would have an area equal to the area observed for the quantity V_n under the t-distribution with $n - 2$ degrees of freedom. The U's can then treated as standard normal random variables and can be used in the regular cusum scheme (Equation 7.1).

7.8 More on the Choice of Cusum Parameters

Hawkins and Olwell (1998) note that "[t]here is often use for a computationally simple approximation to the ARL" (p. 157). Existing approximations, discussed further in the next section, typically approximate the ARL for prespecified values of h and k. A particularly well-known approximation is that given by Siegmund (1985) (see Equation 7.2). Ultimately, such approximations must be solved numerically by trying different values of h to find the particular threshold value that yields the desired ARL. Siegmund's approximation has been useful; its numerical solution has been described by Woodall and Adams (1993), and this has been further modified and used by Alwan (2000). Although such numerical solution is not difficult to carry out, it is interesting to ask whether it is possible to approximate the threshold h directly in terms of k and ARL_0. Despite the fact that alternative methods for finding h are available, a direct and accurate expression for the threshold is potentially useful. Ultimately, the choice of method for determining h is an individual one. For those not wishing to solve the appropriate integral equation, for those who do not find tables complete enough, and for those who do not find nomograms readily available, a simple approximation for h may provide an attractive alternative.

In this section we give expressions that allow direct approximation of the threshold h for given values of k and ARL_0. In addition, we provide formulas for approximating k and h as a function of specified values for the in-control and out-of-control values of ARL. For more details, see Rogerson (2006b).

7.8.1 Approximations for the Critical Threshold h for Given Choices of k and the In-Control ARL_0

When the data are from a standard normal distribution, an approximation for the average run length is (Siegmund 1985)

$$\text{ARL} \approx \frac{e^{-2\theta b} + 2\theta b - 1}{2\theta^2} \tag{7.23}$$

where $\theta = \delta - k$, with δ defined as the size of the change, and k is the reference value. In addition, $b = h + 1.166$, where h is the critical threshold associated with a given cusum scheme. When the process is in control, $\delta = 0$, and the average time until a false alarm is approximately

$$\text{ARL}_0 \approx \frac{e^{2bk} - 2bk - 1}{2k^2} \tag{7.24}$$

In the special and common case where we wish to minimize the time needed to detect a one standard deviation change in the process variable, the optimal value of k is ½, and Equation 7.24 reduces to

$$\text{ARL}_0 \approx 2(e^b - b - 1) \tag{7.25}$$

Typically, one wishes to find the value of b (and hence h) that is associated with some desired ARL_0. However, the value of b corresponding to a specified ARL can only be found by solving Equation 7.24 numerically; it is not possible to solve the equation directly for b. Woodall and Adams (1993) provide an iterative method for solving Equation 7.24 for b given a desired ARL_0. Alwan (2000) modifies and uses this iterative approach. Specifically, Alwan finds the next estimate of h, say h_n, as a function of the previous estimate h_{n-1} as follows:

$$h_n = h_{n-1} - \frac{e^{2k(h_{n-1}+1.166)} - 2k(h_{n-1}+1.1166) - 1 - 2k^2 \text{ARL}_0}{2k e^{2k(h_{n-1}+1.166)} - 2k} \tag{7.26}$$

Despite not being able to solve Equation 7.25 for b directly, it is nevertheless possible to derive a direct approximation for b using the Lambert W function (for more detail see Corless et al. 1996, and for its numerical evaluation, Chapeau-Blondeau and Monir 2002). Using the first two terms of a series expansion of the function, one may derive the following, when $k = \frac{1}{2}$ (Rogerson 2006b):

$$b \approx \ln\left(\frac{ARL_0}{2} + 1\right) \qquad (7.27)$$

Using the first three terms results in the somewhat more accurate approximation:

$$b \approx \frac{ARL_0 + 4}{ARL_0 + 2} \ln\left(\frac{ARL_0}{2} + 1\right) \qquad (7.28)$$

Equations 7.27 and 7.28 provide a direct means of approximating the threshold, $h = b - 1.166$, for a desired ARL_0. In the more general case, when k is not necessarily equal to $\frac{1}{2}$, using the first two terms of the expansion leads to

$$b \approx \frac{\ln(1 + 2k^2\,ARL_0)}{2k} \qquad (7.29)$$

whereas use of the first three terms yields the more accurate

$$b \approx \left(\frac{2k^2 ARL_0 + 2}{2k^2 ARL_0 + 1}\right) \frac{\ln(1 + 2k^2 ARL_0)}{2k} \qquad (7.30)$$

Equations 7.29 and 7.30 are general expressions that allow calculation of $h = b - 1.166$ for prespecified values of k and ARL_0, with Equation 7.30 being more accurate than Equation 7.29. The use of more than three terms of Equation 7.24 will result in a slightly more accurate approximation of the solution to Siegmund's equation; however, the "cost" of such inclusion is the loss of simple expressions such as Equation 7.29 and Equation 7.30 that give direct solutions for the threshold parameter.

Alwan uses an iterative method suggested by Woodall and Adams to solve Siegmund's approximation. He finds $h = 4.93$ for the case where $ARL_0 = 250$ and $k = 0.3399$. Direct application of Equation 7.30 also yields $h = 4.93$ and avoids the need for iterative, numerical solutions. The simpler Equation 7.29 yields $h = 4.83$.

Equation 7.29 will generally be accurate if $0.4 \le k \le 0.8$; Equation 7.30 will generally be accurate for $k \le 1$. More generally, accuracy depends upon the quantity $k^2 ARL_0$; in particular, the accuracy of Equation 7.30 begins to deteriorate when the quantity $k^2 ARL_0 < 1$. Thus k should be greater than $1/\sqrt{ARL_0}$ to ensure that Equation 7.30 is accurate.

7.8.2 Approximations for the Critical Threshold *h* for Given Choices of *k* and the Out-of-Control ARL₁

Suppose that a value of k is chosen, and the process deviates from the target by $2k$; the cusum procedure will detect the change as quickly as possible because the choice of k equal to one-half of the process deviation is optimal with regard to detection time. Although the previous section focused on determination of the threshold value h to achieve a given in-control ARL_0, it is also of great interest to determine what value of h is required to achieve a desired time-to-detection, out-of-control ARL_1, for a prespecified process deviation of $2k$ (Table 7.5). Such a question would

TABLE 7.5

Comparison of Desired and Actual ARL Values When k and ARL_1 Are Prespecified

Desired ARL_1	k	h	ARL_0	Actual ARL_1
2	0.1	4.03	39.4	19.7
	0.2	1.73	12.8	5.9
	0.3	1.10	8.6	3.5
	0.4	0.88	8.0	2.7
	0.5	0.83	9.1	2.4
	0.6	0.87	11.5	2.2
	0.8	1.06	23.8	2.1
	1.0	1.33	67.4	2.1
	1.2	1.65	265.2	2.1
3	0.1	4.13	41.3	20.3
	0.2	1.93	15.1	6.6
	0.3	1.40	11.7	4.2
	0.4	1.28	12.8	3.5
	0.5	1.47	20.2	3.4
	0.6	1.47	26.4	3.1
	0.8	1.86	90.5	3.1
	1.0	2.33	511.1	3.1
	1.2	2.85	4856	3.1
5	0.1	4.33	45.2	21.6
	0.2	2.33	20.6	8.1
	0.3	2.00	20.9	5.8
	0.4	2.08	30.6	5.3
	0.5	2.33	56.6	5.1
	0.6	2.67	128.5	5.1
	0.8	3.46	1235	5.1
	1.0	4.33	28299	5.1
	1.2	5.25	1541229	5.1

(Continued)

TABLE 7.5 (CONTINUED)

Comparison of Desired and Actual ARL Values When k and ARL_1 Are Prespecified

Desired ARL_1	k	h	ARL_0	Actual ARL_1
7	0.1	4.53	49.3	23.0
	0.2	2.73	27.4	9.6
	0.3	2.60	35.0	7.6
	0.4	2.88	66.2	7.2
	0.5	3.33	167.7	7.1
	0.6	3.87	565.7	7.1
	0.8	5.06	16048	7.1
	1.0	6.33	1545736	7.1
10	0.1	4.83	56.0	25.1
	0.2	3.33	40.6	12.1
	0.3	3.50	70.1	10.4
	0.4	4.08	191.2	10.1
	0.5	4.83	786.6	10.1
	0.6	5.67	4978	10.1
	0.8	7.46	747035	10.1
20	0.1	5.83	82.7	32.3
	0.2	5.33	123.2	20.9
	0.3	6.50	520.6	20.1
	0.4	8.08	5065	20.1
	0.5	9.83	118807	20.1
	0.6	11.67	6684217	20.0

arise when there was a particular need to detect a change within a pre-specified average amount of time, and where there was less concern with the rate of false alarms. Although this approach is interesting and important in its own right, it is also of particular use in the next section where we investigate the values of h and k that should be chosen to achieve the desired values of ARL_0 and ARL_1.

We begin by substituting $\delta = 2k$ in Siegmund's Equation 7.23; this leads to

$$ARL_1 = \frac{e^{-2kb} + 2bk - 1}{2k^2} \tag{7.31}$$

Solving this for b yields the following approximation:

$$b \approx k\ ARL_1 + \frac{1}{2k} \tag{7.32}$$

Note that when $k = \frac{1}{2}$, this simplifies to

$$b \approx ARL_1/2 + 1 \qquad (7.33)$$

The accuracy of the formula depends upon the quantity $2k^2ARL_1$. In particular, reasonable accuracy is achieved when $2k^2ARL_1 > 1$. For very small values of k, the actual time taken to detect a change in the mean is greater than desired.

7.8.3   The Choice of *k* and *h* for Desired Values of ARL_0 and ARL_1

In some circumstances, the analyst's sense of how quickly any deviation from the process mean should be detected may be better than the sense of how large the process deviation will be. In such situations it will be of interest to determine the values of k and h for desired values of ARL_0 and ARL_1 (Table 7.6). For instance, the term *out-of-control* does not necessarily

TABLE 7.6

Comparison of Desired and Actual ARL Values When ARL_0 and ARL_1 Are Prespecified

Desired ARL_0	Desired ARL_1	k	h (Equation 7.10)	Actual ARL_0	Actual ARL_1
100	2	1.00	1.50	93	2.2
	3	0.76	2.00	97	3.4
	7	0.40	3.35	100	8.3
	10	0.29	4.11	100	12.6
	20	0.10	6.16	93	34.8
250	2	1.14	1.69	860	2.2
	3	0.88	2.23	238	3.3
	7	0.50	3.70	248	7.4
	10	0.39	4.48	248	11.3
	20	0.22	6.40	249	24.2
	40	0.10	9.24	251	59.7
500	2	1.23	1.82	441	2.2
	3	0.96	2.39	473	3.2
	7	0.56	3.96	494	7.6
	10	0.45	4.80	499	10.9
	20	0.28	6.82	498	22.5
	40	0.16	9.53	501	48.4
1000	2	1.31	1.94	860	2.2
	5	0.76	3.348	976	5.3
	10	0.49	5.11	988	10.7
	20	0.32	7.27	1001	21.7
	40	0.20	10.13	1004	44.9

always correspond to a particular magnitude of deviation from the in-control mean, and the choice of the reference parameter k may be difficult to make (even though it is often customary to simply choose $k = \frac{1}{2}$, to minimize the time taken to detect a change of one standard deviation in the mean). As an alternative to initially specifying k (where the value of ARL_1 is determined as a consequence), the value of ARL_1 may be chosen as the average number of observations within which it is important to determine whether the process has changed. Here k is determined as a consequence of the choice of ARL_1; the resulting value of k indicates one-half of the size of the change that can be detected with the choice of ARL_1. Of course, the choice of ARL_1 should be "reasonable." Although it is clearly desirable that ARL_1 be very low, too low a choice will result in a high value of k (because only very large process changes could be detected so quickly).

Here we find approximations for k and h given the choice of ARL_0 and ARL_1. We may find an expression for k by equating the right-hand sides of Equations 7.29 and 7.32 (although Equation 7.30 is more accurate than Equation 7.29, equating (7.30) and 7.32 does not readily lead to a relatively simple expression for k). Equating, and solving for k, yields

$$k \approx \sqrt{\frac{(\ln(-\ln v) - \ln v)}{2ARL_1} - \frac{1}{2ARL_0}} \tag{7.34}$$

where

$$v = \frac{ARL_1}{ARL_0} e^{1 - ARL_1/ARL_0} \tag{7.35}$$

This value of k will be optimal in finding an out-of-control mean that deviates by $2k$ from the in-control mean; in addition, the average time-to-detection will be ARL_1. As is usual when using a scheme with k, any deviations from the in-control mean that are greater than $2k$ would be detected more quickly, and deviations smaller than $2k$ would take longer to detect. Thus, in prespecifying ARL_1, the analyst determines the minimal deviation (i.e., $2k$) that may be detected that quickly.

After k is determined, b (and therefore h) may be found from Equation 7.30. To illustrate, suppose we choose $ARL_0 = 500$ and $ARL_1 = 7$. Then Equation 7.34 gives $k = 0.563$. Once k is determined, we find $h = b - 1.166$ via Equation 7.30. This gives $h = 3.96$. These parameters are close to the values of $k = 0.6$ and $h = 3.8$ found by Rossi et al. (1999) using nomograms. More importantly, these parameters lead to *actual* in-control and out-of-control run lengths of 494 and 7.6, respectively, via simulation, and these are close to the desired values of 500 and 7. With this scheme, any out-of-control mean that is equal to $2k = 1.126$ will be found with an average run length of about seven; larger out-of-control means will of course be found more quickly.

For a second example, consider $ARL_0 = 500$ and $ARL_1 = 3$. Equation 7.34 yields $k = 0.960$. Using this in Equation 7.30 leads to $h = 2.39$. These parameters are not far from the values of $k = 1.04$ and $h = 2.26$ found by Rossi et al. using nomograms; the *actual* in-control and out-of-control ARLs using $h = 2.39$ and $k = 0.96$ are 473 and 3.2, respectively.

When ARL_0 and ARL_1 are prespecified, Equations 7.34 and 7.30 will provide k and h values that are associated with actual average run lengths that are generally close to the prespecified values. If the derived value of k is in the neighborhood of one or greater, then the actual ARL_0 will be smaller than desired. Likewise, if the derived value of k is smaller than about 0.2, the actual ARL_1 will be greater than desired. One should therefore proceed with this approach only if the derived value of k is in the range (0.2, 1).

7.9 Other Methods for Temporal Surveillance

The exponentially weighted moving average (EWMA) chart was introduced by Roberts (1959) and is discussed further by Hunter (1986) and by Lucas and Saccucci (1990). It is based upon the quantities

$$z_t = (1 - \lambda)z_{t-1} + \lambda x_t \tag{7.36}$$

where x_t is the observation at time t, and λ is a parameter that dictates the importance of dated information. An alarm is signaled at the first time when the value of z_t exceeds a time-varying threshold that over time reaches an asymptotic limit. In the special case of $\lambda = 1$, only current information is used, and the method is identical to the Shewhart chart.

The Shiryaev–Roberts method, based upon contributions from Shiryaev (1963) and Roberts (1966), can be derived as a special case of a likelihood ratio method with a noninformative prior distribution on the time of the changepoint (Frisen and Sonesson 2005). This approach minimizes the expected time until an alarm following a change.

Many other approaches to temporal surveillance exist; these range from simple calculations of historical limits that are empirically based upon recent data to sophisticated use of time series analysis. These are reviewed by Farrington and Beale (1998), and more recently by Le Strat (2005).

8

Spatial Surveillance and the Monitoring of Global Statistics

8.1 Brief Overview of the Development of Methods for Spatial Surveillance

The last chapter introduced methods of statistical process control and paid particular attention to cumulative sum methods as a way to monitor a quantity over time, with the objective of detecting changes in the mean as quickly as possible. In this chapter we begin to address the question of how these methods may be adapted for use with spatial data, where interest lies in the quick detection of change in spatial patterns.

The principles of quality control, and the cumulative sum statistic in particular, have previously been applied to the evaluation of spatial clustering in a limited number of studies. One of the early applications of spatial surveillance was in the area of integrated circuit (IC) fabrication (Collica, Ramirez, and Taam 1996; Hansen, Nair, and Friedman 1997). Hundreds of ICs, or chips, are manufactured simultaneously on silicon wafers, and it is important to monitor the spatial pattern of defects on wafers so that flaws in the fabrication may be detected quickly. Both research teams describe how statistical process methods for monitoring may be used with a spatial join count statistic to carry out spatial surveillance.

Although temporal surveillance has been well developed for health applications (e.g., Chen 1978), there have been fewer developments aimed specifically at spatial surveillance. However, this is changing rapidly, and methods designed for the quick detection of spatial trends include those described by Raubertas (1989), Rogerson (1997, 2001b), Kulldorff (2001), and Kleinman, Lazarus, and Platt (2004). Farrington and Beale (1998) provide a good discussion of some of the issues and difficulties associated with surveillance; Lawson (2001) has also devoted a chapter of his book on spatial methods in epidemiology to surveillance.

Increasingly, there is interest in the prospective spatial surveillance of new data as they become available. Here, the objective is to detect new geographic patterns (e.g., emergent clusters) as quickly as possible. Particularly in light of the perceived threat of bioterrorism, there has been a spate of recent interest in the development of geographic surveillance systems that can detect changes in spatial patterns of disease. Indeed, a recent national conference was devoted entirely to the subject of syndromic surveillance (National Syndromic Surveillance Conference 2002). Efforts in this direction have also been summarized in several articles, some of which have appeared in the popular press (Bunk 2002, Walsh 2002). More general goals of monitoring geographic patterns of public health for outbreaks of, for example, food poisoning and water-borne illnesses, are also garnering increasing attention.

Raubertas (1989) was one of the first to outline how statistical approaches to spatial surveillance could be developed, and he did so in the context of disease surveillance. Raubertas employed cumulative sum (cusum) methods to suggest how disease monitoring for a particular region within a study area could be carried out. Monitoring is based on forming a weighted sum of the number of cases occurring both in the region of interest and in the surrounding regions. The weights define the spatial structure of the alternative hypothesis, and these should therefore be defined as closely as possible according to the spatial structure of relative risks that might be expected in any cluster centered on that location. The weights, for example, might decline as the distance from the region of interest increases. For each time period, the weighted sum of observations is compared with expectations, and deviations are cumulated over time; if these cumulated deviations exceed a prespecified threshold, an alarm is sounded, signaling a possible increase in disease in the vicinity of the region of interest. This is tantamount to monitoring a local statistic over time. Raubertas notes some of the complications that arise when one wishes to monitor several regions simultaneously because there will be correlation in the monitoring statistics obtained for regions that are close to one another (because they will have shared neighborhoods).

In addition to monitoring local statistics, early development of spatial surveillance focused upon the development of methods for monitoring global spatial statistics. One strategy for this is to place a global spatial statistic within the cusum monitoring framework as demonstrated by Rogerson (1997), who uses cusum methods to monitor temporal changes in Tango's (1995) global spatial statistic. First, Tango's statistic is computed for a particular set of observations (e.g., disease cases) aggregated into a set of subregions. Each time a new case is observed, Tango's statistic is updated and compared with a *conditional* expectation of the statistic under the null hypothesis of no raised incidence in any subregion. The expectation used here is conditional because it is calculated based upon the given, previous set of observations, that is, before the latest case is added to the set. A conditional variance is also calculated in the same way. Then, the expectations and variance are used to convert the updated value of Tango's statistic into a z-score, indicating how likely or unlikely it is to observe the updated value of the statistic given the previous status of the subregional system. Finally, these z-scores are used in a cusum framework and an alarm is sounded, indicating a significant change in the global statistic if the z-scores cumulate sufficiently. In principle, other global spatial statistics could also be used in this same framework. Rogerson and Sun (2001) show how a similar approach may be used to monitor changes in the nearest-neighbor statistic, whereas Rogerson (2001b) monitors changes in the space–time Knox statistic as new observations are collected.

Continuing with the tripartite classification suggested by Besag and Newell (1991), spatial monitoring has also been carried out by modifying the scan-type statistics used for the detection of clustering when the size and location of clusters is not known a priori. For example, it is possible to monitor many regions, or the local statistics associated with many regions, simultaneously, with adjustments for the multiple testing that occurs as a result of the multiplicity of regions. This is discussed further in Chapter 10.

As one example from this category, Kulldorff (2001) has extended his spatial scan statistic (see Chapter 5, Section 5.7) to the case of prospective disease surveillance, by considering the likelihood of the observed number of events in space–time cylinders, where the vertical axis or height of the cylinder represents time, and the horizontal plane represents a region and its surrounding neighborhood, under the null hypothesis. In this space–time scan statistic, the original circular scanning windows are replaced by cylinders with time on the vertical axis, where the top of the cylinder represents the most recent time period. Then, the maximum likelihood ratio concept is simply generalized to search for space–time clusters as the cylinders grow vertically with the progression of time. At each time period, the likelihood of the most interesting cylinder (i.e., the one with the highest likelihood ratio) is compared with the likelihood of the most interesting cylinder generated from many simulations of the null hypothesis. The popularity of the method has been aided by freely available software, *SaTScan*, which was briefly introduced in Chapter 1, Section 1.7.1.4. However, no account is made of the multiple testing that occurs from period to period. The nature of Kulldorff's approach suggests that it may also be fruitful to combine the spatial scan statistic with the cumulative sum method proposed here.

Yet another approach to spatial surveillance is demonstrated by Kleinman, Lazarus, and Platt (2004), who model the count of cases in a small region using covariates in a generalized linear mixed model (Breslow and Clayton 1993) for a historical period. In particular, they use a logistic equation to model the probability that a particular individual is a case. Next, they use the coefficient estimates to derive the expected probability that an individual becomes a case during the next time period. Statistical significance is achieved if the observed count of cases is unlikely to have occurred, and this is assessed by using a binomial distribution based upon the number of individuals and the predicted probability resulting from the model. They apply their model to data on the daily counts of the number of patients visiting medical clinics in the Boston area for lower respiratory infection. Although they have data for a large number of census tracts, their method is not particularly spatial as it would miss the scenario where a cluster of adjacent regions was exhibiting evidence of increasing rates.

Other approaches to spatial surveillance include distance-based methods (e.g., see Forsberg et al. 2005) and perspectives that adopt more of a model-based than a statistical hypothesis testing perspective (Lawson 2005).

The organization of the last part of this book is along the same lines as the first part, where methods for retrospective detection of geographic clustering were discussed in order according to Besag and Newell's tripartite classification. This third part of the book consists of four chapters. In this chapter we have introduced the concepts of prospective surveillance in a context that is not explicitly spatial. In the remainder of this chapter, we start the discussion on spatial surveillance by first examining the monitoring of global statistics. Chapter 9 covers the monitoring of local statistics. We also provide in Chapter 9 a treatment of the monitoring of scan-type statistics, where multiple regions are monitored simultaneously. Chapter 10 discusses yet other possible approaches to the statistical surveillance of geographic clustering. Chapter 11 covers associated tests for cluster detection and surveillance.

8.2   Introduction to Monitoring Global Spatial Statistics

We saw in Chapter 3 that a global spatial statistic is a single numerical summary measure of a spatial pattern. Typically, such statistics are calculated for a single map, and often a hypothesis test is then carried out to determine whether there is a statistically significant departure from either spatial randomness or some set of predefined spatial expectations.

There may be many instances, however, when global spatial statistics are available for repeated periods of time, and there may be interest in determining whether there are temporal changes in global spatial autocorrelation or spatial dependence. For instance, one may wish to discern whether a series of maps displays an increasing or decreasing tendency for similar cancer rates to cluster near to one another. Crime analysts may wish to detect whether the geographic pattern of crime is becoming increasingly clustered, which in turn may lead to a reallocation of patrol and enforcement efforts.

It is possible to carry out repeated tests of significance—one for each map—but the issue of multiple hypothesis testing must be borne in mind. Even when the null hypothesis of no map pattern is true, repeated testing with, for example, a Type I error probability of .05 will lead to an average of one false declaration for every 20 time periods tested. One general approach for making adjustments for the repeated testing of spatial statistics is to create a Shewhart surveillance environment. Suppose, for example, that we have a sequence of maps and an associated temporal sequence of values for Moran's I. We can declare the system to be out-of-control (i.e., where we reject the null hypothesis of no spatial autocorrelation) when the value of I exceeds the $1/ARL_0^{th}$ percentage point of the normal approximation that is commonly used for evaluating the statistical significance of I.

Standard methods for spatial pattern analysis are not applicable to surveillance problems, where the objective is the quick detection of changes in spatial patterns, and new "prospective" methods thus need to be developed. Especially in the field of health research, there is increasing recognition of the need for methods and systems that are designed for the prospective detection of spatial clusters of disease. Farrington and Beale (1998), for example, have recently argued that "in contrast to an extensive statistical literature on the *retrospective* detection of temporal and spatial clusters of disease, rather less attention has been devoted to the statistical problem of *prospective* outbreak detection" (p. 98). For health research aimed at prospective surveillance, observations are processed sequentially, and an ongoing surveillance effort is made to detect emerging clusters as new observations are reported and combined into prior data.

In this chapter, we introduce cusum methods as a way for monitoring the global statistics; they address the problem of multiple testing and are optimal for finding temporal changes in the global spatial statistics. There are two general approaches for the temporal surveillance of global statistics. In the first, global statistics such as Moran's I for each time period are first standardized and then used in a cusum monitoring scheme. In the second approach, the new information available during each time period is first used to update a global statistic. The updated global statistic is then compared with its expectation, conditional upon both the previous period's global statistic and the presumption that the process is in control (i.e., the spatial pattern of disease risk is uniform). The standardized differences between observations and expectations are next embedded within the cusum scheme. Examples of this second approach include updating of Tango, Knox, and nearest-neighbor statistics (Rogerson 1997, 2001b; Rogerson and Sun 2001). On a historical note, their examples were, chronologically, among the earlier suggestions made for spatial surveillance.

In addition to the rapid detection of the onset of spatial clustering in an otherwise random spatial pattern of disease, other applications are possible. Detecting change in patterns of urban crime, for example, could prove useful in reallocating patrol beats. Kulldorff (1999) suggests a number of other application areas. Many of these would benefit not only from the retrospective analysis of spatial clustering but also a system of monitoring changes in patterns over time. For example, ecologists have interest in the spatial pattern of species of flora and fauna, but also of interest is whether and how particular geographic distributions change over time. Transportation engineers may be interested in not only the current spatial distribution of traffic accidents but also how the distribution pattern changes over time as new developments on and along a highway system occur.

8.3   Cumulative Sum Methods and Global Spatial Statistics That Are Observed Periodically

8.3.1   Moran's *I* and Getis' *G**

In this section, we will use two global statistics—Moran's *I* and Getis' *G*—to illustrate how a temporal sequence of such global statistics may be monitored, using the cusum methods reviewed in Chapter 7.

A straightforward approach to combining global spatial statistics and statistical process control is to simply cumulate the spatial statistics (after they have been converted to a distribution that is approximately standard normal) for each time period, as they are observed. Thus, the cumulative sums for the standardized *I* and *G* statistics at time *t* are, respectively,

$$S_t = \max(0, S_{t-1} + Z[I_t] - k)$$

$$S_t = \max(0, S_{t-1} + Z[G_t] - k)$$

(8.1)

where $Z[I]$ and $Z[G]$ are standardized *I* and *G* statistics, and where *k* will usually be chosen to be equal to ½ (as described in this chapter), implying that the cumulative scheme is optimized to find increases in the *z*-scores equal in magnitude to one standard deviation. Note that monitoring increases in this context implies that we wish to detect increasing concentration of higher-than-average values (measured by the *G* statistic) or increasing positive spatial autocorrelation (measured by the *I* statistic).

The independence assumption required for the cusum translates here into the assumption that the statistical measure of spatial association in one time period is independent of the measure in other time periods.

To summarize, the primary focus of this analysis is temporal; the objective is to find deviations from expectations as quickly as possible. The "variable" that is being subject to this temporal analysis is a global spatial statistic. Hence, the intent is to find as quickly as possible, through

---

* This section was written jointly with Gyoungju Lee.

the temporal cusum analysis, a change in the global measure of spatial pattern.

8.3.1.1 Example: Breast Cancer Mortality
in the Northeastern United States

The data set on annual breast cancer mortality is used here for illustrative purposes only. One somewhat undesirable feature of this example is the great length of each time period; when a spatial change occurs, it will still take several years (or time periods) to find it.

Two different spatial scales were used to determine binary spatial weights—24.2 miles and 47 miles. More specifically, county pairs were included in the calculation of Moran's I and Getis' G if their centroids were within these distances of one another. These two scales correspond approximately to: (1) using adjacent counties (because the centroids of adjacent counties in the northeastern United States are approximately 25 to 30 miles apart), and (2) in addition, using counties that are adjacent to adjacent counties, respectively.

The cusums in Equation 8.1 were used to monitor changes in global spatial association in breast cancer mortality. We chose a value of $ARL_0 = 100$, which yields a threshold of $h = 2.84$. This corresponds to one false alarm every 100 years. As the distribution of average run lengths when the process is in control (i.e., there is no significant increase in spatial association) is approximately exponential, this implies that the probability of signaling in a given year is approximately $1/100 = .01$. Consequently, the probability of no signal over the 30-year monitoring period is approximately $.99^{30} = .74$, and thus the probability of a false alarm is approximately $1 - .74 = .26$.

We implemented monitoring in two ways: (1) using the usual cusum, which assumes a baseline value of $z = 0$, corresponding here to a baseline where the global statistics are monitored for deviations from their expected values in a random pattern, and (2) using a self-starting cusum, where the baseline value of z is found from early observations of the data. This latter approach is useful if, for example, there is spatial association in the early observations, and one wishes to monitor future deviations from it.

Table 8.1 displays the results of the cumulative sum for Moran's I. For Table 8.1(a), Moran's I values were calculated using the z-scores created from observed and expected numbers of breast cancer cases, assuming a Poisson distribution under the null hypothesis of no raised incidence; hence, $z = (Y_{obs} - Y_{exp})/\sqrt{Y_{exp}}$. The expected values are found by indirect standardization, assuming that national group-specific mortality rates apply to population groups in each county. Table 8.1(b) displays the results of using Moran's I based upon the standardized mortality ratio (SMR), which is computed as the ratio of observed to expected numbers of deaths due to breast cancer.

The cusum in Table 8.1(a) exceeds its critical threshold for the smaller geographic scale in 1988; the cusum for the larger scale is exceeded in 1990. Furthermore, the cusum continues to trend upward after these dates. These findings indicate that the map of z-scores displays increasing spatial association. Results for the maps of SMRs shown in Table 8.1(b) reveal a different picture; the cusum exceeds its critical value only during 1973–1974 for the smaller geographic scale, and during 1973–1976 for the

TABLE 8.1

Cusum Analysis of Moran's I for the Breast Cancer Mortality Data for the Northeastern United States

	(a) Based on z-Scores					
	128,000 ft (24.2 mi)			248,000 ft (47 mi)		
Year	z-Value	Cusum	Self-Starting Cusum	z-Value	Cusum	Self-Starting Cusum
1968	0.3665	0	—	0.9074	0.4074	—
1969	0.4929	0	—	0.1837	0.0911	—
1970	−1.2681	0	0	−1.2904	0	0
1971	0.5351	0.0351	0	0.6001	0.1001	0
1972	1.0944	0.6295	0.43	0.387	0	0
1973	0.353	0.4824	0.04	1.4558	0.9558	0.68
1974	0.0946	0.077	0	0.272	0.7279	0.09
1975	−0.8461	0	0	−1.1885	0	0
1976	−0.6717	0	0	−0.4792	0	0
1977	0.7849	0.2849	0.39	−1.9225	0	0
1978	0.6134	0.3983	0.51	0.2721	0	0
1979	1.9898	1.8881	2.08	0.0329	0	0
1980	−0.3655	1.0226	0.89	0.3842	0	0
1981	−0.0555	0.4671	0.06	−0.4636	0	0
1982	−0.2854	0	0	−0.5032	0	0
1983	0.8059	0.3059	0.21	−0.0934	0	0
1984	1.2431	1.049	0.88	1.3709	0.8709	1.05
1985	0.1761	0.7251	0.25	1.1964	1.5672	1.79
1986	−1.3347	0	0	−1.2666	0	0
1987	3.2192	2.7192	2.48	−0.3704	0	0
1988	2.0418	4.261	3.45	1.1289	0.6289	0.68
1989	0.9888	4.7498	3.44	2.4443	2.5732	2.5
1990	2.8734	7.1233	4.98	1.5697	3.6429	3.29
1991	2.5431	9.1663	6.06	−0.3189	2.824	2.32
1992	3.2107	11.877	7.51	3.7514	6.0754	4.8
1993	5.8066	17.1836	10.35	3.8468	9.4221	6.84
1994	5.134	21.8176	12.23	0.9617	9.8839	6.69
1995	0.4862	21.8039	11.4	2.2838	11.6676	7.44
1996	2.1539	23.4577	11.5	2.4828	13.6505	8.27
1997	3.4086	26.3663	12.27	1.0668	14.2173	8.08
1998	0.7675	26.6338	11.54	−0.9462	12.7711	6.5

TABLE 8.1 (CONTINUED)

Cusum Analysis of Moran's *I* for the Breast Cancer Mortality Data for the Northeastern United States

	(b) Based on SMR					
	128,000 ft (24.2 mi)			248,000 ft (47 mi)		
Year	z-Value	Cusum	Self-Starting Cusum	z-Value	Cusum	Self-Starting Cusum
1968	−1.3104	0	—	−0.4689	0	—
1969	1.1616	0.6616	—	0.5282	0.0282	—
1970	−0.6517	0	0	−0.7208	0	0
1971	1.6421	1.1421	0.48	2.6235	2.1235	1.35
1972	1.1007	1.7428	0.49	1.2325	2.8559	1.24
1973	2.1804	3.4231	1.08	3.0927	5.4486	2.1
1974	0.4341	3.3572	0.42	0.0606	5.0093	1.06
1975	−0.5231	2.3341	0	−0.9009	3.6083	0
1976	0.2392	2.0733	0	0.0398	3.1481	0
1977	0.6117	2.1851	0	−0.6277	2.0204	0
1978	0.1613	1.8463	0	0.0369	1.5573	0
1979	−0.2221	1.1242	0	−0.2084	0.8489	0
1980	0.0228	0.6471	0	1.3372	1.6861	0.18
1981	−1.3409	0	0	−0.4558	0.7304	0
1982	0.0781	0	0	−1.3636	0	0
1983	−0.1573	0	0	0.6534	0.1534	0
1984	−0.879	0	0	0.4152	0.0686	0
1985	−0.498	0	0	0.5746	0.1432	0
1986	−1.2918	0	0	0.2896	0	0
1987	3.2581	2.7581	2.29	0.8763	0.3763	0
1988	−0.3304	1.9277	1.36	−0.3551	0	0
1989	−1.0026	0.4252	0	0.1015	0	0
1990	0.0578	0	0	−0.4932	0	0
1991	1.3666	0.8666	0.54	−0.0713	0	0
1992	0.9908	1.3574	0.72	1.1779	0.6779	0.35
1993	1.7853	2.6427	1.54	1.4105	1.5884	0.89
1994	−0.0015	2.1412	0.82	0.613	1.7013	0.65
1995	−1.1582	0.483	0	0.2229	1.4243	0.03
1996	0.3594	0.3424	0	−0.3164	0.6079	0
1997	−0.1936	0	0	0.033	0.1408	0
1998	−0.7645	0	0	0.0785	0	0

Note: The standardized mortality ratio (SMR) is the ratio of observed to expected deaths.

larger scale. Assessing the sequence of SMR maps using Getis' *G* reveals a qualitatively similar picture as shown in Table 8.2, although the critical threshold is never exceeded.

Figure 8.1 displays values of local Moran's *I* for selected years. For 1993, notice that there is clustering of high local Moran statistics based on z-scores in and near the New York City metropolitan area. These

TABLE 8.2

Cusum Analysis of Getis' *G* Statistic Based on SMR for the Breast Cancer
Mortality Data for the Northeastern United States

	128,000 ft (24.2 mi)			248,000 ft (47 mi)		
Year	z-Value	Cusum	Self-Starting Cusum	z-Value	Cusum	Self-Starting Cusum
1968	0.6182	0.1182	—	0.2993	0	—
1969	−0.4115	0	—	−1.7444	0	—
1970	1.1386	0.6386	0.25	1.4071	0.9071	0.27
1971	0.7395	0.878	0.03	1.3397	1.7468	0.38
1972	0.1479	0.526	0	0.9836	2.2304	0.25
1973	−2.3391	0	0	−2.182	0	0
1974	−0.0046	0	0	0.4465	0	0
1975	1.3111	0.8111	0.49	1.2806	0.7806	0.22
1976	0.2898	0.6009	0.1	−0.0539	0.2267	0
1977	−0.4016	0	0	−1.2534	0	0
1978	−1.0752	0	0	−0.1836	0	0
1979	−0.7508	0	0	−1.122	0	0
1980	−1.5878	0	0	−1.9819	0	0
1981	−1.1281	0	0	−0.9487	0	0
1982	0.3032	0	0	−0.433	0	0
1983	−1.6744	0	0	−1.5778	0	0
1984	0.2571	0	0	0.4314	0	0.11
1985	−1.4226	0	0	−1.1586	0	0
1986	−0.3518	0	0	−0.2793	0	0
1987	−0.9815	0	0	−1.1056	0	0
1988	−1.6279	0	0	−1.7763	0	0
1989	−1.3116	0	0	−1.8922	0	0
1990	0.0445	0	0	0.3235	0	0.2
1991	−0.1477	0	0	−0.4889	0	0
1992	−0.8143	0	0	−0.243	0	0
1993	0.013	0	0	−0.1836	0	0
1994	0.0434	0	0	0.8089	0.3089	0.63
1995	−0.318	0	0	−0.7689	0	0
1996	1.0567	0.5567	1.05	1.0217	0.5217	0.8
1997	1.3062	1.3629	2.25	1.4224	1.4441	1.88
1998	−1.8862	0	0.18	−1.624	0	0.26

Note: The standardized mortality ratio (SMR) is the ratio of observed to expected deaths.

contribute to the high global statistic (relative to previous years). This feature is notably absent from the corresponding 1993 local Moran map for
SMR values. This is a high population density area; although the SMRs
are not far above the average SMR, the associated z-scores are well above
average; observed mortality is statistically well above expected, despite
the fact that the relative risk is not raised much.

Comparing the z-score and SMR maps in Figure 8.1 for 1973 reveals that
Moran's *I* is relatively higher and more significant for the SMR map than

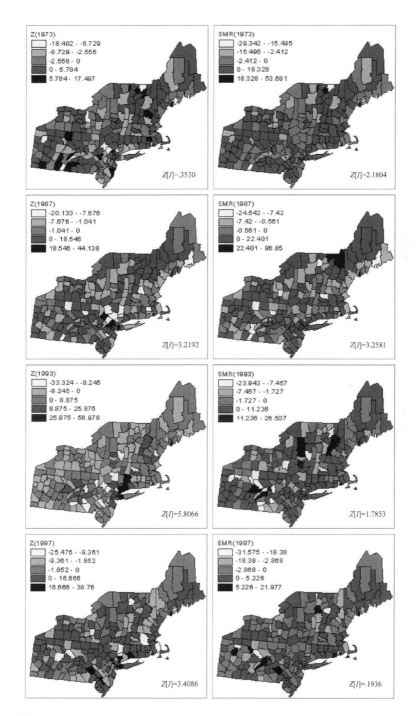

FIGURE 8.1
Map of local Moran's *I* for *z*-scores and SMRs (1973, 1987, 1993, and 1997).

for the z-map. Note that areas where SMRs are above average, and areas where they are below average, each cluster together. However, these are areas that are more rural than urban. Although the SMRs deviate from the average and are clustered because the clusters are located primarily in rural regions, the statistical significance as reflected in the z-scores is not high. Hence, the corresponding areas on the map of z-scores are not particularly significant, and do not cluster.

Tables 8.1 and 8.2 also display the results for a self-starting cusum. These results are broadly similar to those already discussed, and alarms signaling change are sent at roughly the same time as in the usual cusum. This implies that a standard assumption of no global spatial association at the start of the period is fairly reasonable for this case study.

In summary, the SMR and z-maps present different pictures of changes in breast cancer mortality. The sequence of SMR maps shows that there is not substantial change over the period being studied, although there is a period during the mid-1970s where slightly more clustering occurs, but primarily in rural regions. The z-map and the associated cusum display more substantial changes. The cusum signals a shift toward more spatial association during the early 1990s. This signaling is due at least in part to a higher clustering of z-scores in and around the New York City region.

8.3.2   Monitoring Chi-Square Statistics

As noted in Chapter 3, a number of global spatial statistics (including the aspatial goodness-of-fit statistic, Tango's statistic, and the spatial chi-square statistic), have distributions that may be approximated by the chi-square distribution. If data are collected periodically, these spatial statistics may be derived periodically, and it is of interest to detect change in them quickly. In particular, we ask here how to use successive global spatial statistics observed over time (and that have chi-square distributions) within a cusum framework.

Hawkins and Olwell (1998) outline a cusum procedure for chi-square random variables. The chi-square distribution is a special case of the gamma distribution; the latter can be written as

$$f(y) = \frac{y^{\alpha-1}e^{-y/\beta}}{\beta^{\alpha}\Gamma(\alpha)} \tag{8.2}$$

where $\Gamma(\alpha)$ is the gamma function, defined as $\Gamma(z) = \int_0^{\infty} t^{z-1}e^{-t}dt$; when z is an integer, $\Gamma(z) = (z - 1)!$.

The chi-square distribution corresponds to choices of $\alpha = (m - 1)/2$ and $\beta = 2$, where $m - 1$ is the number of degrees of freedom. The mean of the chi-square distribution is equal to $\alpha\beta = m$. The β is a scale parameter, and we wish to monitor for changes from it to some higher value, say β_1.

Suppose that, for the aspatial chi-square goodness-of-fit statistic, we wish to minimize the time it takes to find a change, where one region's z-score increases from a mean of zero to a mean of one, whereas the remaining $m - 1$ regions have compensating declines in the means of their z-scores, from zero to $-1/(m - 1)$. Under this alternative, the expected value of the (chi-square distributed) global statistic will increase from $m - 1$ to $\mu_1 = m - 1 + \frac{m}{m-1}$. This suggests that the shape parameter β should increase from 2 to

$$\beta_1 = 2\left[1 + \frac{m}{(m-1)^2}\right] \tag{8.3}$$

Then, using results from Hawkins and Olwell (p. 143), the following cusum can be used to detect rapidly an upward shift in the global statistic:

$$S_{i,t} = \max\left(0, S_{i,t-1} + \chi_t^2 - k\right) \tag{8.4}$$

where

$$k = \frac{(m-1)\{\ln(\beta_1) - \ln(2)\}}{1 - 2\beta_1^{-1}} \tag{8.5}$$

An approximation that can be used here is based on the fact that we are looking for a change in the expectation of the goodness-of-fit statistic from $m - 1$ under the null hypothesis to roughly m under this particular alternative hypothesis. Thus k turns out to be roughly halfway between these two values, at approximately $m - 0.5$.

Srivastava (1997) suggests the following approximation for the threshold, h:

$$\text{ARL}_0 \approx \frac{e^{ah_1} - ah_1 - 1}{a(m-1)(k-1)} \tag{8.6}$$

where $a = (\mu_1 - 1)/2\mu_1$ and $h_1 = h + \sqrt{2}ck$. The constant, c, is a function of the number of degrees of freedom $(m - 1)$ (see Table 8.3).

8.3.2.1   *Illustration*

Suppose that for a system of $m = 7$ regions, we periodically collect data on the number of observed and expected cases in each region. We wish to monitor the chi-square goodness-of-fit statistic. Equation 8.3 implies that

TABLE 8.3

Constant (c) as a Function of Degrees
of Freedom ($m - 1$) for Determination
of Cusum Threshold

Degrees of Freedom ($m - 1$)	c
1	1.4874
2	1.3333
3	1.2785
4	1.2490
5	1.2339
6	1.2225
7	1.2144
8	1.2081
9	1.2035
10	1.1996
11	1.1965
12	1.1939

Source: Srivastava, M.S. (1997).

Note: For larger degrees of freedom, the constant asymptotically approaches the value of 1.166 used for the normal distribution.

$\beta_1 = 2.39$ and Equation 8.5 implies that $k = 6.55$ (which is close to $m - 0.5$, as noted previously). Further, suppose that we wish to monitor successive global statistics, with a desired ARL_0 of 100. We first find $\mu_1 = 7.17$ and $a = 0.43$. Using Equation 8.6 with either trial and error or numerical methods yields $h_1 = 16.91$. Finally, $c = 1.223$, implying that $h = 16.91 - \sqrt{2}(1.223)(6.55) = 5.58$.

8.4 CUSUM Methods and Global Spatial Statistics That Are Updated Periodically

In some cases, data used for calculation of the global statistic will include information from past time periods. For instance, one may accumulate the number of cases found in each subregion of a study region and compute a global statistic for the entire study region. Thus when a new case is observed, the global statistic is updated using a new set of accumulated counts, which includes all cases used for the computation of the previous global statistics and the newly observed case. In this scenario, it is important to devise a monitoring scheme that separates out the new information and uses only the new information in the cusum for each period. For example, Rogerson (1997) adopts Tango's (1995) statistic as a measure of spatial pattern (in principle, other spatial statistics can also be used). Suppose, for example, that Tango's statistic is calculated for a particular set of observations. Then, using the prespecified multinomial probabilities that a case will fall into a particular subregion, one computes the probability that a newly computed Tango's statistic, based upon both the previous observations and one more observation, will take on a particular value. This leads to, for the null hypothesis, calculation of the expected value and variance of the Tango statistic after inclusion of the next observation, conditional upon the current value of the statistic. In turn, the expected value and variance may be used to convert the Tango statistic that is observed after the next observation into a z-score. Finally, these z-scores are used in a cusum framework. Rogerson and Sun (2001) show how a similar approach may be used to monitor changes in the nearest-neighbor statistic, whereas Rogerson (2001b) monitors changes in the space–time Knox statistic as new observations are collected.

These ideas are now discussed in more detail. We begin in the next subsection with a description of how we can modify Tango's statistics associated with general and focused tests of clustering and use them sequentially to form cumulative sum statistics useful for detecting change in spatial patterns. The procedure is first illustrated with simulated data comparable to that used by Tango (1995). The section that follows illustrates the application of the cusum statistic to data on Burkitt's lymphoma in Uganda.

8.4.1 Spatial Surveillance Using Tango's Test for General Clustering

Tango's global statistic is described for use in a retrospective sense in Chapter 3, Section 3.8. Our present interest is in monitoring the location of new cases as they occur, with the objective of detecting emerging clusters shortly after they occur. It would be inappropriate simply to use Tango's C_G statistic after each new observation because of the problem of testing multiple hypotheses over time. Instead, we develop a cusum approach based upon the observed values of C_G (computed after the first i observations), and the corresponding expected values, conditional upon the observed value of C_G that is based upon the first $i - 1$ observations.

Let $\mathbf{r}_{i-1}$ denote the $m \times 1$ vector containing the proportion of observed cases in each region after the first $i - 1$ observations, where m is the number of regions in the regional system being monitored. Similarly, let $\mathbf{p}$ be an $m \times 1$ vector containing the expected proportions of cases in each region. Also denote Tango's C_G statistic based on the first i observations as

$C_{G,i}$. Then, the expectation of $C_{G,i}$, conditional upon $C_{G,i-1}$, is

$$E[C_{G,i} \mid C_{G,i-1}] = \mathbf{p}'\mathbf{u} \tag{8.7}$$

where $\mathbf{u}$ is an $m \times 1$ vector containing as element k:

$$u_k = (\mathbf{r}_{i-1}(k) - \mathbf{p})' \mathbf{A} \ (\mathbf{r}_{i-1}(k) - \mathbf{p}) \tag{8.8}$$

where $\mathbf{r}_{i-1}(k)$ denotes the proportion of cases in each region, given that case i is located in region k when $\mathbf{r}_{i-1}$ is the vector containing observed proportions after the first $i - 1$ observations. In addition, $\mathbf{A}$ is an $m \times m$ matrix containing elements w_{jk}, which measure the closeness (or connectivity) between regions j and k. The conditional variance is

$$\sigma^2_{C_{G,i} \mid C_{G,i-1}} = \mathbf{p}'(\text{diag } \mathbf{uu}') - (\mathbf{p}'\mathbf{u})^2 \tag{8.9}$$

where "diag" refers to the $m \times 1$ vector created from the diagonal of $\mathbf{uu}'$.
     The cusum is based upon the quantities

$$Z_i = \frac{C_{G,i} - E[C_{G,i} \mid C_{G,i-1}]}{\sigma_{C_{G,i} \mid C_{G,i-1}}} \tag{8.10}$$

Large, positive values of Z_i signal substantial deviations of Tango's statistic from its conditional expectation. If these positive deviations cumulate sufficiently, a cluster alarm sounds. Although the Z_i's have mean 0 and variance 1, they do not have a normal distribution. To satisfy the assumption of normality, we take observations of Z_i in batches of size n, and cumulate the quantity $Z_{(n)}$ defined as the mean of the Z_i's in each batch. The number of successive values should be sufficiently large to confer normality, but it is desirable to keep n as small as possible to facilitate quick detection of changes. The $Z_{(n)}$ have a mean of zero and a standard deviation of $1/\sqrt{n}$.

8.4.1.1   Illustration

To illustrate, we follow Tango's example and choose the centroids of $m = 100$ hypothetical regions at random within the square defined by x- and y-coordinates ranging from 0 to 40. The populations for each region are assigned a random number from the distribution $\exp(N(0,1))$. Figure 8.2 depicts the study region, with the radius of each circle drawn proportional to population size.
     Tests of normality on simulated data for this scenario confirmed that, under the null hypothesis of no emerging clusters, the $Z_{(n)}$'s come from a normal distribution when the batch size n is as small as 4, and so we take $n = 4$. To make the cusum scheme somewhat comparable to Tango's

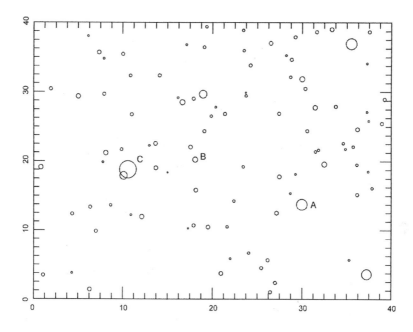

FIGURE 8.2
A hypothetical study region for cumulative sum (cusum) surveillance with Tango's C_G statistic.

retrospective test on $N = 100$ cases of disease, let us choose h so that the probability of a false alarm during the first 100 observations (or 25 batches) is approximately $\alpha = 0.05$. Although most cusum tables give ARLs, we need here the *distribution* of the run length. We can choose the appropriate value of h by using the fact that the run length distribution is approximately exponential, with parameter equal to $\mu = 1/\text{ARL}$. We seek h such that the probability of an alarm in the first 25 batches (or 100 observations) is .05. Thus $\Pr(RL < 25) \approx 1 - e^{-25\mu} = .05$, which implies that $\mu = .00205$. The ARL is $1/\mu = 487.4$ batches, or 1950 observations. Use of this value with an ARL table (e.g., Table 7.1), and with the usual choice of $k = \frac{1}{2}$, implies that the value of h is approximately 4.3. In situations like this one, where the quantity being monitored does not have a standard deviation of one, the parameters h and k are adjusted by multiplying by the standard deviation. Here we have $h = 4.3/\sqrt{4} = 2.15$ and $k = 0.5/\sqrt{4} = 1/4$ as the adjusted parameters to be used in the cusum. The ultimate choice of h lies in the hands of the analyst; one could, for example, choose a higher value if one desired fewer false alarms.

The use of the cusum approach also relies upon the assumption that the sequence of Z values is independent; that is, no serial correlation in the values is permitted. To test the validity of this assumption, we carried out a Monte Carlo simulation. We generated 250 successive values of $2Z_{(4)}$ under the null hypothesis, and the autocorrelation function revealed no significant serial dependence at any temporal lag.

We next simulated alternative scenarios:

1. The relative risk (RR) of the disease is equal to 1 in each of the $m = 100$ regions. This corresponds to the null hypothesis.

2. RR = 3 for a given region, together with an exponential decline in RR with distance (more specifically, $\exp(-d/\tau)$, where d is the distance from the given region and $\tau = 5$) from the region. Regions having RR < 1.5 after this assignment of relative risks were reassigned RR = 1. This is similar to Tango's designation of an area comprised of a small number of regions affected outward from the center of the disease. We then used Points A, B, and C in Figure 8.2 separately and successively as designated centers of simulated disease.

Table 8.4 displays the results of 1000 repetitions of each of the four scenarios, and compares the cusum approach with Tango's C_G statistic used retrospectively after $N = 100$ observations with $\alpha = 0.05$. When the simulated disease rate is equal in each region, Tango's statistic was significant in 4.6% of the cases (near the nominal value of 5%). The rate at which the cusum statistic signals a false alarm is similar (6.8%); recall that we chose h and k to achieve an approximate false-alarm rate of 5% during the first 100 observations.

When there is an elevated relative risk of disease in and around a given region, the cusum statistic can detect this with an average run length that depends in part upon the size (i.e., population) of the region as well as the size and number of nearby regions. Point C is the largest population center and has another large region close by; not surprisingly, with an elevated relative risk of disease around point C, the ARL is relatively short (88.3 observations). Note that the distribution of run lengths is positively skewed; the median run length is somewhat lower than the ARL.

TABLE 8.4

Results of Cumulative Sum Surveillance with Tango's C_G Statistic on Simulated Data

Simulated Clustering	Probability of Rejecting H_0 after $N = 100$ Observations		ARL	Median Run Length	ARL | RL < 100
	Tango (T_G)	Cusum			
None	.046	.068			
Around Point A	.370	.364	160.0	140	60.2
Around Point B	.404	.397	147.7	120	61.4
Around Point C	.736	.676	88.3	76	57.1

Note, also, that the power of the cusum statistic, in terms of the likelihood of signaling an alarm within the first 100 observations, is slightly lower than that for Tango's retrospective statistic. The "cost" of surveillance and quick detection of changes in pattern is slightly lower power in comparison with a retrospective test that uses a given N. It is important to emphasize once again that, in the usual application of tests for clustering, we use only one value of N, and surveillance or repetition of the test over time encounters the problem of testing multiple hypotheses. The strength of surveillance lies not so much in its statistical power as in its timeliness.

8.4.1.2 Example: Burkitt's Lymphoma in Uganda

Williams et al. (1978) provide data on the locations of children (and a few young adults) with Burkitt's lymphoma, along with dates of onset and dates of diagnosis, for the period 1961–1975 in Uganda. Figure 8.3 portrays the 10-region study area; readers interested in the data should refer to the

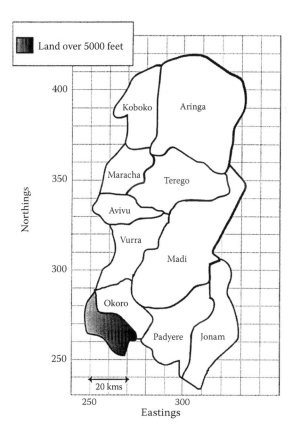

FIGURE 8.3
The study region for the Burkitt's lymphoma data provided in Williams et al. (1978).

original paper, where the data appear in an appendix. Williams et al. use chi-square tests and Knox's test to find space–time clustering during the periods 1961–1965 and 1972–1973. Bailey and Gatrell (1995, pp. 124–125), in a reanalysis of the data, also find space–time clustering.

Here, we employ the cusum version of Tango's test to monitor the development of spatial pattern in lymphoma cases, relative to the background population, for the 188 cases where location and date of diagnosis are available. Using the available 1968 mid-year population, surveillance for general clustering commenced in January 1961. We used values of $n = 4$, $k = \frac{1}{2}$, and $h = 4.2$. The Shapiro–Wilk statistic failed to reject the assumption that the 188/4 = 47 batches of $Z_{(4)}$ come from a normal distribution.

To give meaningful weight to the importance of surrounding regions, we chose a value of $\tau = 20$ in $w_{jk} = \exp(-d_{jk}/\tau)$. This assigns a closeness measure of 1 for a region's closeness to itself, values of about 0.2 for adjacent, surrounding regions, and values less than 0.1 for other, more distant regions. In this illustration, values lower than about $\tau = 10$ give almost no weight to the importance of other, surrounding regions—the spatial nature of any clustering, therefore, is undetectable, and the test is essentially similar to a nonspatial chi-square test.

The cusum test first signaled the development of significant clustering in March 1967, after observation of 56 cases. Until then there was no significant departure detected in the spatial pattern of lymphoma cases from the mid-1968 population distribution. The second column of Table 8.5(a) shows the observed and expected proportions of cases in each region, among the first 56 observations. Region 4 (Terego) has approximately twice the number of expected cases, given its population size, and this was an important contributor to the alarm. This excess continued during the period after the alarm (observations 57–104), as shown in the third column. Values of τ between 1 and about 15 lead to a signal after the first 60 cases; values of τ ranging from 15–30 lead to a cluster alarm after the first 56 cases; and τ ranging from about 30–100 (corresponding to a decreasingly severe decline with distance) all yield an alarm after the first 40 cases. This appears to confirm the results cited earlier, indicating relative robustness with respect to the choice of τ.

In quality control applications, it is common to reset the cusum to zero following an alarm (the cusum in principle should only be reset when it is possible to assign a cause to, and understand, the change). After resetting the cusum in March 1967, surveillance for general clustering again commenced; the next alarm occurred in December 1970, after 48 more observations. The third column of Table 8.5(a) shows that the eastern regions 2, 4, and 6 (Aringa, Terego, and Madi) all have more cases than expected under the null hypothesis. Surveillance commenced once again following the December 1970 signal, this time using estimated 1972 populations (derived with the regional growth rates given by Williams et al. 1978). Additional cluster signals occur after observations 148 (in May 1973) and 172

TABLE 8.5

Comparison of Expected Proportion of Cases in Each Region
with Observed Proportion

(a) For the First 104 Observations with Expectation Based on 1968 Population

| | Proportion of Cases in Each Region | | |
| | Observed Proportion of Cases | | Expected Proportion of Cases |
Region	Obs. 1–56	Obs. 57–104	Based on 1968 Population
1 Koboko	.0357	.0208	.0644
2 Aringa	.1429	.2083	.0999
3 Maracha	.1429	.0833	.1060
4 Terego	.2143	.2083	.1013
5 Ayivu	.1429	.0417	.1342
6 Madi	.0893	.1458	.0807
7 Vurra	.0714	.0625	.0622
8 Okoro	.0179	.0208	.1365
9 Padyere	.0714	.1042	.1308
10 Jonam	.0714	.1042	.0841

(b) For the Latter 68 Observations with Expectation Based on 1973 Population

| | Proportion of Cases in Each Region | | |
| | Observed Proportion of Cases | | Expected Proportion of Cases |
Region	Obs. 105–148	Obs. 149–172	Based on 1973 Population
1 Koboko	.0455	.0833	.0685
2 Aringa	.1591	.1250	.0958
3 Maracha	.1364	.1667	.0969
4 Terego	.0682	.2917	.0954
5 Ayivu	.2955	.2083	.1283
6 Madi	.0682	.0000	.0846
7 Vurra	.1136	.0833	.0573
8 Okoro	.0455	.0000	.1418
9 Padyere	.0227	.0417	.1359
10 Jonam	.0455	.0000	.0950

Note: Columns may not sum to 1.000 because of rounding off.

(in February 1975). Table 8.5(b) reveals that the relatively large number of cases in regions 5 and 7 (Ayivu and Vurra) contributed to the 1973 signal, and a relatively large number of cases in regions 3, 4, and 5 contributed to the 1975 signal. Figure 8.4 shows how the cusum statistic changed over time.

Many possible causes may have led to the cluster signals. For example, changes in the distribution of the population at risk may have occurred. To evaluate this possibility, the monitoring was repeated, this time using estimated 1972 populations from the start of surveillance in 1961. The results were nearly identical, suggesting that changes in the background population are not causing the detected clustering. It is, of course, possible that changes in the population distribution of *children* may have contributed

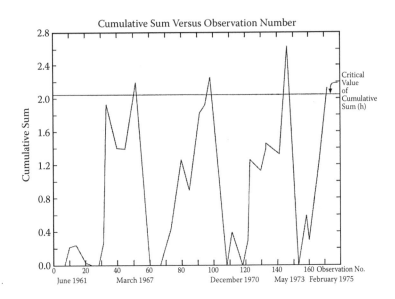

FIGURE 8.4
Cusum chart using Tango's C_G statistic for the Burkitt's lymphoma data.

to the signal, but this would require fairly significant and unlikely shifts in the distribution of children. Data on regional populations by age are unavailable.

Williams et al. (1978) also note that regional variations exist in the ability to detect cases. They discuss, and then essentially dismiss, the prospect that *changes* in the relative abilities to detect cases occurred in such a manner to cause clustering. They suggest that real clustering may result from the possibility that Burkitt's lymphoma occurs following chronic and severe infection with malaria. Whatever the cause, the value of the cusum approach lies in its ability to monitor and detect clusters.

8.4.1.3   Discussion

The test statistic described relies on successive computations of Tango's statistic, as new observations become available. Although Tango's statistic has many desirable characteristics, there is, of course, no particular reason to adopt this specific statistic, and one could develop fruitfully similar cusum statistics for other measures of pattern; a similar approach is taken in the following subsections for the Knox and nearest-neighbor statistics.

One potential shortcoming of the approach described earlier for general clustering is that the statistic is based on an assessment of change over the entire study area. Thus, there is reliance on global, as opposed to local, statistics; local areas of change might remain concealed by lumping results together from many surrounding areas that have not experienced change.

8.4.2 A Cusum Method Based upon the Knox Statistic: Monitoring Point Patterns for the Development of Space–Time Clusters

In this section, cusum methods are adapted for use with Knox's space–time statistic. The result is a procedure for the rapid detection of any emergent space–time interaction for a set of sequentially monitored point events. The approach relies upon a "local" Knox statistic that is useful in retrospective analyses to detect when and where space–time interaction occurs. In the following subsections, the distribution of the local Knox statistic under the null hypothesis of no space–time interaction is first derived. The results associated with the local Knox test are then employed to demonstrate how cusum methods may be used in conjunction with the global Knox statistic calculated sequentially over time to monitor potential changes in space–time interaction as new data are collected. The retrospective local statistic and the prospective cusum monitoring method are illustrated using the data on Burkitt's lymphoma in Uganda, also used in the previous section.

8.4.2.1 A Local Knox Statistic

When the Knox statistic discussed in Chapter 6, Section 6.2 proves significant in retrospective analyses, the observations that contribute most to the space–time interaction are not indicated. In this sense, the Knox statistic is *global*; significant space–time interaction exists in the data, but the specific times and locations that interact remain unidentified. In this subsection we describe a local Knox statistic that may be defined for each individual observation. A significant local statistic for an observation implies that the observation has more links that are close both in space and in time than would be expected for that observation.

For a dataset consisting of n observations, each observation may be compared with the $n - 1$ other observations. Let $n_s(i)$ be the number of observations that are close to observation i in space, $n_t(i)$ be the number that are close in time, and $n_{st}(i)$ be the number that are close to i in both space and time. Define the observed value of the local Knox statistic $N_{st}(i)$ to be equal to $n_{st}(i)$. The local statistic $n_{st}(i)$ satisfies Anselin's (1995) requirement that the local statistics sum to a constant multiple of the global statistic because

$$\sum_{i=1}^{n} n_{st}(i) = 2n_{st} \qquad (8.11)$$

To determine the distribution of the local Knox statistic for observation i under the null hypothesis, we begin with the assumption that each random permutation of the times across the fixed spatial locations is equally likely. Let $j = 1, \ldots, n$ be the index of the time value assigned to point i, and define $n_t^j(i)$ as the number of points close in time to point i when point i is assigned the j-th value of time. For a given permutation of times across the spatial locations, the distribution of $N_{st}(i)$ is hypergeometric, with parameters $n - 1$, $n_s(i)$, and $n_t^j(i)$. The null distribution of $N_{st}(i)$ is given as a weighted sum of these hypergeometric distributions, where the equal weights reflect the equally likely permutations:

$$\Pr(N_{st}(i) = n_{st}(i)) = \frac{1}{n} \frac{\sum_{j=1}^{n} \binom{n_t^j(i)}{n_{st}} \binom{n-1-n_t^j(i)}{n_s(i)-n_{st}}}{\binom{n-1}{n_s(i)}} \tag{8.12}$$

where the quantity $\binom{a}{b}$ is equal to $a!/[b!(a-b)!]$

To carry out a test of significance, the exceedance probability may be written as

$$\Pr(N_{st}(i) \geq n_{st}(i)) = \sum_{k=n_{st}}^{\max_j n_t^j(i)} \frac{1}{n} \frac{\sum_{j=1}^{n} \binom{n_t^j(i)}{k} \binom{n-1-n_t^j(i)}{n_s(i)-k}}{\binom{n-1}{n_s(i)}} \tag{8.13}$$

where the vacuous terms (e.g., where $k > n_t^j(i)$) are set equal to zero. This may be rewritten as

$$\Pr(N_{st}(i) \geq n_{st}(i)) = \frac{1}{n} \sum_{j=1}^{n} \left[1 - F\{n_{st}(i) - 1\}; n - 1, n_s(i), n_t^j(i) \right] \tag{8.14}$$

where $F()$ refers to the cumulative distribution function of the hypergeometric random variable. For large values of the parameters, Berry and Mielke (1998) suggest a rapid computation method for the cumulative distribution function.

Alternatively, a normal approximation may be used, where

$$E[N_{st}(i)] = \frac{\sum_{j=1}^{n} n_t^j(i) n_s(i)}{n(n-1)}$$

$$V[N_{st}(i)] = \frac{\sum_{j=1}^{n} n_t^j(i) n_s(i) \{n-1-n_t^j(i)\}\{n-1-n_s(i)\}}{n(n-1)^2(n-2)} \tag{8.15}$$

These may be simplified to

$$E[N_{st}(i)] = \frac{2n_t n_s(i)}{n(n-1)}$$

$$V[N_{st}(i)] = \frac{\{2(n-2)n_t - 2n_{2t}\}n_s(i)\{n-1-n_s(i)\}}{n(n-1)^2(n-2)}$$

(8.16)

where n_{2t} and n_t retain their earlier definitions given in Chapter 6, Section 6.2 (recall that the latter is the number of space–time pairs within time t_0 of one another, that is, $\sum_{i=1}^{n} n_t(i) = 2n_t$). A significance test for $n_{st}(i)$ may be constructed by taking the quantity

$$z_i = \frac{n_{st}(i) - E[N_{st}(i)]}{\sqrt{V[N_{st}(i)]}}$$

(8.17)

to have, approximately, a standard normal distribution under the null hypothesis of no space–time interaction at point i. The probability of exceeding this z-score z_i should ideally be identical to the probability of obtaining a value of the random variable $N_{st}(i)$ equal to or more extreme than the one observed. That is, we desire

$$1 - \Phi(z_i) = \Pr(N_{st}(i) \geq n_{st}(i))$$

(8.18)

where $\Phi(z_i)$ denotes the cumulative distribution function of a standard normal distribution, evaluated at z_i. Because $N_{st}(i)$ is a discrete variable, an adjustment to the numerator of z_i is required to make the quantities on the left-hand and right-hand sides of Equation 8.18 approximately equal. A correction factor accounting for the discreteness of the variable may be employed by subtracting 0.5 from the numerator of the z-score previously defined by Equation 8.17 to ensure that the two are approximately equal. The adjusted z-statistic is therefore

$$z_i = \frac{n_{st}(i) - E[N_{st}(i)] - 0.5}{\sqrt{V[N_{st}(i)]}}$$

(8.19)

When the global Knox statistic is insignificant, the local statistic may prove useful in uncovering local, space–time "hot spots," although an adjustment (e.g., a Bonferroni adjustment) should be made if multiple local tests are carried out. When the global Knox statistic is significant, the local Knox statistic can be useful in finding the observations that contribute to the global significance. It is also possible that the local Knox statistics could reveal that the significant global statistic was due to a small number of outlying observations.

In the next section, we make use of these results for local statistics to develop a prospective monitoring system for space–time interaction.

8.4.2.2 A Method for Monitoring Changes in Space–Time Interaction

With each new observation, we wish to assess whether there is increased evidence for space–time interaction. After each new observation, therefore, we are interested in comparing any observed increase in the Knox statistic with what would have been expected under the null hypothesis.

Suppose that information on $i - 1$ cases has been observed to date. For observation i, let $n_s(i)$, $n_t(i)$, and $n_{st}(i)$ be defined as the number of links to the previous $i - 1$ observations that are close together in space, in time, and in space and time, respectively. We will make use of this relationship between the time and space coordinates of the newest observation and the space–time coordinates of previous observations.

Monitoring of the Knox statistic may be carried out by comparing the value of the Knox statistic after observation i (denoted K_i) with the value that would be expected under the null hypothesis, conditional upon the value of the Knox statistic after observation $i - 1$ and the observed values of $n_s(i)$ and $n_t(i)$. Specifically, define the z-score

$$z_i = \frac{K_i - E[K_i \mid K_{i-1}]}{\sqrt{V[K_i \mid K_{i-1}]}} \tag{8.20}$$

The expectation and variance may be given as

$$E[K_i \mid K_{i-1}] = \sum_{x=0}^{\min\{n_s(i),\max[n_t^j(i)]\}} (K_{i-1} + x)\Pr(N_{st}(i) = x)$$

$$\tag{8.21}$$

$$V[K_i \mid K_{i-1}] = \sum_{x=0}^{\min\{n_s(i),\max[n_t^j(i)]\}} (K_{i-1} + x)^2 \Pr(N_{st}(i) = x) - \{E[K_i \mid K_{i-1}]\}^2$$

where $\Pr(N_{st}(i) = x)$ is the conditional probability that the addition of observation i to the data set contributes x additional pairs of points that are close in space and time, given $n_s(i)$ and $n_t(i)$. Using the results of the previous section, the variable $N_{st}(i)$ is the weighted sum of hypergeometric variables:

$$\Pr(N_{st}(i) = x) = \frac{1}{i} \frac{\sum_{j=1}^{i} \binom{n_t^j(i)}{x} \binom{i-1-n_t^j(i)}{n_s(i)-x}}{\binom{i-1}{n_s(i)}} \tag{8.22}$$

where $n_t^j(i)$ is the number of points that are close to observation i in time, when observation i is assigned the j-th indexed value of time.

The information contained in this comparison of K_i with its conditional expectation is due solely to the specific contribution of observation i to the Knox statistic. Rogerson (2001b) shows that the z-score defined in Equation 8.20 is identical to that associated with the local Knox statistic for observation i, after i observations have been taken:

$$z_i = \frac{n_{st}(i) - E[N_{st}(i)]}{\sqrt{V[N_{st}(i)]}} \tag{8.23}$$

where the expectation and variance are equal to

$$E[N_{st}(i)] = \frac{n_s(i)\sum_{j=1}^{i} n_t^j(i)}{i(i-1)}$$

$$V[N_{st}(i)] = \frac{n_s(i)\{i-1-n_s(i)\}\sum_{j=1}^{i} n_t^j(i)\{i-1-n_t^j(i)\}}{i(i-1)^2(i-2)} \tag{8.24}$$

With the addition of the correction factor used due to the discreteness of $n_{st}(i)$, the z-score used in the cusum is

$$z_i = \frac{n_{st}(i) - E[N_{st}(i)] - 0.5}{\sqrt{V[N_{st}(i)]}} \tag{8.25}$$

The cusum, following observation i, is then calculated as

$$S_i = \max(0, S_{i-1} + z_i - k) \tag{8.26}$$

The cumulative sum will exceed its critical value (i.e., $S_i > h$) if observations displaying space–time interaction begin to accumulate. Such observations would have the characteristic that, among their linkages to observations that are recent (i.e., close in time), there would be more linkages than expected to observations that were also close together in space.

8.4.2.3 Example: Burkitt's Lymphoma in Uganda

Using the data on Burkitt's lymphoma described in Section 8.4.1.2, monitoring of the Knox statistic was carried out for the same combinations of critical time windows and spatial distances (t_0 and s_0) as used by

Williams et al. (1978). The values of t_0 were 60, 90, 120, 180, and 360 days. The values of s_0 were 2.5, 5, 10, 20, and 40 km. The parameter h was set at 5.5, corresponding to an ARL_0 of approximately 1560. Because the average run length distribution is approximately exponential under the null hypothesis, this choice implies that the probability of a false alarm during the study period is approximately 0.1 (because $1 - e^{-188/1560} \approx 0.1$; as a reminder the number of cases is 188).

Table 8.6 summarizes the results. Significant space–time interactions are found at critical space and time units that are broadly similar to the original findings. The table also indicates when the original space–time clustering signals would have been noted if monitoring of observations had taken place as new data were received.

The original analysis by Williams et al. did not find space–time clustering during the 1971–1975 period, but *did* find clustering when the data set was limited to those with onset during the period 1972–1973. A continuous monitoring approach not only has the advantage of finding clusters in a timely fashion but also eliminates the need to break retrospective data into rather arbitrary temporal slices in the search for significance.

Figure 8.5 provides an example of how the cumulative sum may be tracked over time for the case $s_0 = 20$ km and $t_0 = 180$ days. In the figure, a total of three space–time clustering signals are indicated. A brief signal is sent at observation 26 in December 1964. A second signal is noted at observation 82 (February 1969), and again this is very brief in duration; that is, it is followed by a period of observations that are not consistent with space–time interaction. The most persistent signal begins at observation 146 (January 1973). This indication of clustering remains until observation 172 in January 1975. An examination of the individual contributions of each observation (that is, the individual z-values associated with each observation) reveals that observations 145–165 (occurring during the period January–November, 1973) were the primary contributors to this signal.

To illustrate the use of the local Knox statistic, we continue with the example using $s_0 = 20$ km and $t_0 = 180$ days. The period 1961–1965 contains 35 observations, and the original analysis by Williams et al. found significant space–time interaction for this period with this set of critical time and distance parameters. Using a one-tailed test with $\alpha = 0.05$ and a Bonferroni adjustment for $n = 35$ tests yields a critical value of $z = 2.98$. Observations 8, 22, 23, 25, 26, 27, and 28 have significant local statistics that exceed this critical value.

8.4.2.4 Summary and Discussion

The Knox statistic has been combined here with a cusum approach to develop a method for the prospective detection of space–time interaction.

TABLE 8.6

Results of the Cusum Surveillance with the Global Knox Test for Different
Critical Space and Time Parameters Applied to the Burkitt's Lymphoma Data

s_0	t_0	Observations at Which "Out-of-Control" Signal Occurs	Date on Which "Out-of-Control" is Signaled
2.5	360	—	
5	360	—	
10	360	54–71, 76, 82 153, 155–56	Dec. 1966; Dec. 1968; Feb. 1969; Apr. 1973; May. 1973
20	360	9, 26–35; 38–83, 148–49, 153–72	Jul. 1962; Dec. 1964; Jan. 1966; Feb. 1973; Apr. 1973
40	360	9, 23, 25–188	Jul. 1962; Aug. 1964 (2)
2.5	180	—	
5	180	—	
10	180	82, 148–49, 153–55	Feb. 1969; Feb. 1973; Apr. 1973;
20	180	26–30, 82, 146–72	Dec. 1964; Feb. 1969; Jan. 1973
40	180	8–14, 21–96, 121–88	Jun. 1962; May 1964; Aug. 1971
2.5	120	—	
5	120	—	
10	120	—	
20	120	41–43, 148, 155–56, 159–68	Mar. 1966; Feb. 1973; Apr. 1973; Jul. 1973
40	120	41–44, 46, 48, 121–42, 144–88	Mar. 1966; Aug. 1966 (2); Aug. 1971; Dec. 1972
2.5	90	141–42	Nov. 1972
5	90	—	
10	90	—	
	90	40–41	Mar. 1966
40	90	41–44, 48, 122–32, 147–88	Mar. 1966; Aug. 1966, Aug. 1971, Jan. 1973
2.5	60	—	
5	60	—	
10	60	—	
20	60	—	
40	60	124–25, 148–78	Sep. 1971; Feb. 1973
30	—	—	
30	—	—	
30	—	—	
30	—	—	
40	30	—	

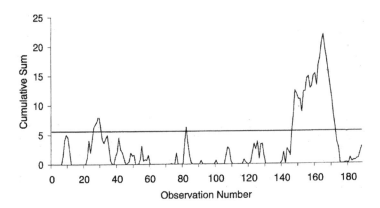

FIGURE 8.5
Cusum chart using the global Knox text for the Burkitt's lymphoma data ($s_0 = 20$ km and $t_0 = 180$ days).

The method makes use of a local Knox statistic, which is also useful in finding those observations that contribute to the global significance of space–time interaction in retrospective studies.

8.4.3 Cusum Method Combined with Nearest-Neighbor Statistic

This section describes a procedure for detecting changes over time in the spatial pattern of point events, combining the (global) nearest-neighbor statistic and cumulative sum methods. It may also be implemented using time windows of differing length to search for any changes in spatial pattern that may occur at particular time scales. The method is illustrated using 1996 arson data from the Buffalo, New York, police department.

8.4.3.1 Monitoring Changes in Point Patterns

There are at least two reasons why it is not desirable to repeat statistical tests that use the nearest-neighbor statistic, which was introduced in Chapter 3, Section 3.2. First, one must account for the fact that an adjustment should be made for the number of tests being carried out. Consider the following simulation. A total of 50 points were successively located in the unit square. Following the location of each point (beginning with the second point because a single point cannot be thought of as a cluster), a nearest-neighbor statistic was calculated and its associated z-statistic was then computed using the means and standard deviations given in

Table 8.7.* This z-score was then compared with the critical value of $z = -1.96$ (corresponding to a one-tailed test with $\alpha = 0.025$, or a two-tailed test with $\alpha = 0.05$). In 23% of the 10,000 replications, the null hypothesis of no clustering was rejected before 50 points had been located in the square. This high percentage is due to more than one hypothesis having been tested.

An adjustment may be used to account for the 49 separate tests being conducted; such an adjustment uses the fact that we want the probability that no significant result has been found after all 49 tests have been carried out to be equal to, say, .975 (i.e., $(1 - x)^{49} = .975$), where x is the probability of rejection in a single test. In this case, solving for x yields $x = .00052$, and the corresponding critical value of z is $z = -3.28$. This adjustment is conservative (because the separate tests are not actually independent); in 10,000 simulations of the null scenario of 50 randomly located points in the unit square, only .6% of the time was the null hypothesis rejected (compared to the nominal value of 2.5%). Less conservative adjustments that account for the correlation between tests are not straightforward to derive.

Just as important, there is a great deal of "inertia" in the nearest-neighbor statistic when it is calculated repeatedly after each new point has been located. If points begin to cluster, the nearest-neighbor statistic may not decline quickly because it will always be based on an *average* of the distances to *all* nearest neighbors, and not just the distance to nearest neighbors for the most recent points. Thus, it may take a long time for changes to appear in the statistic.

8.4.3.2 A Cusum Approach for the Nearest-Neighbor Statistic

We combine the nearest-neighbor and cusum methods as follows. At each stage in the evolution of an observed point pattern (e.g., when $t - 1$ points have been observed to date), we locate a point at random on the map, and the distance from this point to its nearest neighbor is calculated. This is repeated a large number of times, and the mean $(\bar{d})$ and

---

* One difficulty in implementing the statistical test described in Section 8.4.3 concerns boundary effects; the nearest neighbor of a point inside the study area may lie outside the study area. To obviate the difficulties caused by boundary effects, a simulation was performed to find critical values of the nearest-neighbor statistic. N points were randomly located in the unit square, and the nearest-neighbor statistic R was calculated. For each value of N, this was repeated 10,000 times. The resulting mean values of R and the standard deviations of R associated with tests for clustering in the unit square are shown in Table 8.7. Note that, in a bounded region such as the square used here, the observed distance between nearest neighbors will be somewhat greater than that expected in a random pattern, yielding a mean value of R slightly greater than one (because distances to near neighbors lying outside the bounded region are discarded).

TABLE 8.7

Simulated Mean and Standard Deviation of the Nearest-Neighbor Distance
Statistic R in a Random Point Pattern in the Unit Square

Number of Points (N)	Mean R	Std. Dev. R	Number of Points	Mean R	Std. Dev. R
2	1.470	.7120	36	1.078	.1016
3	1.347	.4924	37	1.076	.1018
4	1.286	.4041	38	1.075	.0989
5	1.248	.3436	39	1.075	.0972
6	1.222	.2994	40	1.076	.0967
7	1.203	.2745	41	1.072	.0950
8	1.188	.2511	42	1.071	.0941
9	1.180	.2311	43	1.070	.0919
10	1.164	.2202	44	1.068	.0917
11	1.152	.2035	45	1.070	.0899
12	1.144	.1954	46	1.067	.0888
13	1.138	.1837	47	1.067	.0874
14	1.135	.1767	48	1.068	.0874
15	1.132	.1700	49	1.065	.0853
16	1.121	.1659	50	1.064	.0853
17	1.118	.1590			
18	1.117	.1521			
19	1.111	.1486			
20	1.111	.1453			
21	1.106	.1384			
22	1.101	.1353			
23	1.100	.1313			
24	1.099	.1283			
25	1.096	.1257			
26	1.094	.1235			
27	1.092	.1197			
28	1.088	.1170			
29	1.085	.1138			
30	1.085	.1133			
31	1.084	.1108			
32	1.084	.1096			
33	1.082	.1067			
34	1.079	.1065			
35	1.080	.1045			

variance (σ_d^2) of the distances from the randomly located points to their
nearest neighbors are found. Then, a z-score is assigned to observation
t as follows:

$$z = \frac{d_{obs} - \bar{d}}{\sigma_d} \tag{8.27}$$

where d_{obs} is the observed distance from point t to its nearest neighbor. These quantities may then be cumulated in a cusum scheme as described by Equation 8.26 to detect departures from randomness in the direction of uniformity. Such a scheme would signal a change when observed distances between neighbors began to exceed the distances expected in a random pattern. To detect departures from randomness in the direction of clustering, one would use

$$S_t = \max(0, S_{t-1} - z - k) \tag{8.28}$$

and again, a signal of change in pattern is sent when $S_t > h$.

Because distances to nearest neighbors do not follow a normal distribution, the assumption of normality, required by the cusum approach, is violated. As described in Section 8.3.1, a solution is to aggregate successive, normalized observations into batches by summing the z-scores within a batch. For example, if we batch together five observations at a time by summing the five consecutive z-scores, we have created a new variable (say $z_{(5)} = \Sigma z_i / 5$). With a batch size given by b, the mean of $z_{(b)}$ will still be equal to zero, and the variance of $z_{(b)}$ is equal to $1/b$. Thus, we may cumulate the quantities $(z_{(b)} - 0)/(1/\sqrt{b})$ in the cusum scheme. Usually, the value of b can be quite small for the assumption of normality to be acceptable; for the simulations in the unit square, a batch size of $b = 3$ was found to be acceptable.

8.4.3.3   Simulations of Clustering in the Unit Square

The simulation scenario described above was modified to generate clustering as follows. After observation $t = 20$, points were located randomly with x- and y-coordinates in the interval $(0, 0.25)$ with probability .2, and located randomly within the entire $(0,1)$ square with probability .8. Sequential use of the nearest-neighbor statistic resulted in detection of clustering in 54.3% of the 10,000 simulations on or before the 50th observation (in 40.3% of the simulations, clusters were correctly detected after observation 20). With the conservative adjustment for multiple testing, clusters were found on or before the 50th observation in only 10.1% of the simulations (in 9.8% of the simulations, clusters were found after observation 20).

Using the combined cusum–nearest-neighbor approach described, clusters were detected in 97.7% of the 10,000 simulations on or before the 50th observation. The mean observation number at which a clustering signal was received was 38.5—a bit more than 18 observations after clustering began. Note the substantial improvement in cluster detection in comparison with the sequential use of the nearest-neighbor test. Sequential use of the nearest-neighbor test is hampered by the inertia associated with the first 20 observations, which follow the null hypothesis of no clustering.

Even after a change in process occurs after observation 20, the nearest-neighbor statistic calculated after subsequent observations will contain information based on the first 20 observations, and hence, it declines only slowly.

8.4.3.4   Example: Application to Crime Analysis and Data from the Buffalo Police Department

Clarke (1980, 1992) and Eck (1995) have argued that criminology has shifted its emphases from offenders to offences and from motivations to opportunities, in which the role of places is increasingly recognized. These transitions have pushed the study of crime from the national, regional, and metropolitan scales down to scales below that of the neighborhood, and they have been facilitated by advances in geographical information system technologies (Taylor, 1997). Such approaches often start from examining such questions as "where are the crimes" instead of "who did it" (Sherman 1995). It has been argued that an approach focusing on the locations of crimes should be more effective than those focusing on offenders, particularly for places that show consistently higher incidences of crime.

Areas of high incidence, or "hot spots," have been receiving growing attention (Green, 1995; Kenney and Forde, 1990; Weisburd and Green, 1995). Crime analysts, in addition to being interested in the identification of existing crime hot spots, are also interested in methods that can quickly detect new, emerging hot spots so that policing efforts can be allocated more efficiently.

Data on the locations and times of 379 arsons were available for 1996 from the Buffalo Police Department (BPD). The data represent actual incidents; the data are to be distinguished from emergency calls for service (which would include false alarms and potentially multiple records for a single incident) and arrest data. Figure 8.6 shows how the nearest-neighbor statistic changes throughout the year for arsons. There appears to be a fairly steady decline in R for arsons during 1996. Because the statistic changes only slowly over time, we next turn to a cusum approach in the next subsection to determine if and when significant changes take place in the underlying geographic pattern.

8.4.3.5   Cusum Approach for Arson Data

The cusum nearest-neighbor approach described in Section 8.4.3.2 was used with the 1996 BPD arson data. Presumably, this approach will be more sensitive in identifying points in time where the pattern has changed, in comparison with the trends shown in Figure 8.6 (which has a great deal of embedded inertia).

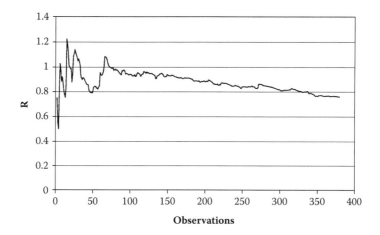

FIGURE 8.6
Nearest-neighbor statistics over time for 1996 arson data for Buffalo, New York.

The first 200 arsons of 1996 (occurring during the period from January to July) were used as a "base" pattern. For this base period, it was determined that the location of each successive point was an average of 0.216 times the expected distance from its nearest neighbor. This yields a baseline measure of clustering that exists in the population; arsons cluster during the base period because population is clustered and because crimes such as arsons may tend to occur more in some areas than in others. The reader should keep in mind that here we are interested in deviations from this baseline amount of clustering, and not in the detection of clustering itself. Recall that an original estimate of expected distance is determined by locating a point at random within the study area, computing the distance to its nearest neighbor, and then repeating this many times; we now wish to scale this expected distance downward in accordance with the observed clustering). Thus, we use

$$z = \frac{d_{obs} - .216\bar{d}}{.216\sigma_d} \tag{8.29}$$

in place of Equation 8.27. An alternative would be to use the baseline locations to estimate a kernel density estimate of arson occurrences. One could then sample from this as a way of generating points to estimate $\bar{d}$ and σ_d (e.g., see Brunsdon 1995). The correlation between successive values of z (i.e., the correlation between z_t and z_{t+1}) was found to be insignificant, and thus the underlying assumption of no serial autocorrelation is satisfied.

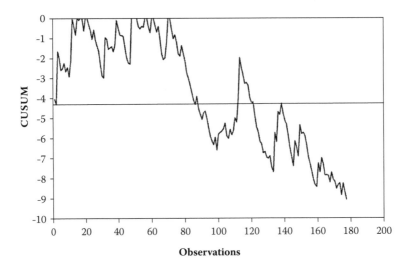

FIGURE 8.7
Cusum for the 1996 arson data. *Note:* Cusum not reset to zero following alarm.

A more complete assessment would also include examination of higher-order correlations such as the correlation between z_t and z_{t+2}.

Surveillance of the pattern then began in August. Values of $k = \frac{1}{2}$ and $h = 4.12$ were used. The value of $h = 4.12$ was arrived at by using Equation 7.4, after choosing an ARL_0 value of 380 (corresponding to approximately one false clustering alarm per year). Figures 8.7 and 8.8 show that the cusum statistic becomes critically negative, indicating clustering relative to the base pattern, after about 80 observations (in October). Note from Figure 8.7, where the cusum was not reset to zero after signaling, that the cusum statistic remains in the critical region for most of the remainder of the year. Changing the definition of the base period from the first 200 observations to either the first 100 or first 150 observations also leads to a cluster signal at this same time. This stability of the signaling with respect to changes in the base period provides reassuring evidence that results are not overly sensitive to minor changes in the definition of the base period. The cusum statistic is commonly reset to zero following a signal (especially in industrial process control, where the change, often a defect, can be noted and the equipment or process appropriately modified). Figure 8.8 demonstrates that, when the cusum is reset to zero, the signal of clustering is given two additional times before the end of the year, indicating that the cause of the increased clustering has persisted.

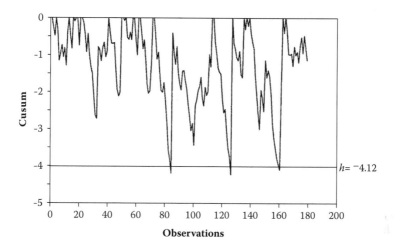

FIGURE 8.8
Cusum for the 1996 arson data. *Note:* Cusum reset to zero following alarm.

Figures 8.9 and 8.10 depict the spatial pattern of arsons during 1996. Figure 8.9 contains 12 black triangles, representing those 1996 arsons that occurred just prior to the first cluster signal, and thus could be considered to have caused it. Figure 8.10 contains triangles, representing those 1996 arsons that followed the first cluster signal. The maps show that the triangles are nearer to neighboring arsons than the dots are to their neighbors—arsons occurring later in the year were more likely to belong to clusters.

A natural question to ask is *why* the pattern has changed. One possibility is that there is seasonal variation in the pattern; because we only had access to data for one year, we were unable to further explore that possibility. Another possibility is that it was the base period that was unusual; perhaps the spatial pattern during the first half of 1996 was *less* clustered and more uniform and spread out than is usual. Again, study of data in adjacent years would help shed additional light on this question.

It is important to note that other types of surveillance may also be desired. For example, we may wish to detect deviations from the base period that occur in the opposite direction of what we have been considering here—namely, distances from new arsons to their nearest neighbors that are *greater* than expected. This would perhaps indicate that arsons were beginning to occur at new locations (which in turn might be the result of geographic displacement following an enforcement effort). Or we may wish to find periods of time where *recent* arsons are located nearer to one another in comparison with some base period. This latter example is treated in the next section.

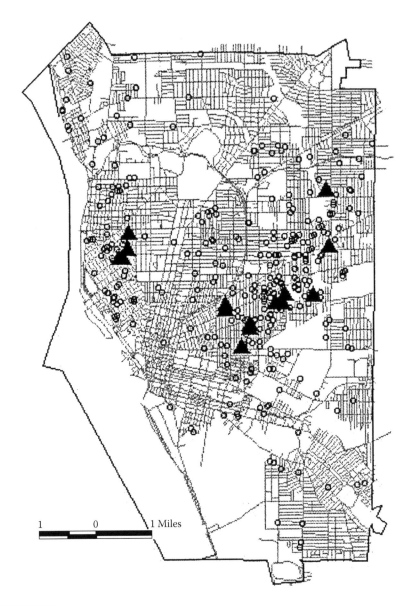

FIGURE 8.9
Spatial distribution of arsons in 1996. Triangles represent arsons leading to cluster signal in early October.

8.4.3.6 *Surveillance Using a Moving Window of Observations*

One of the characteristics of the surveillance method as described to this point is that the nearest-neighbor distance for a newly observed point is calculated as the minimum of the distance to *all previous observations*.

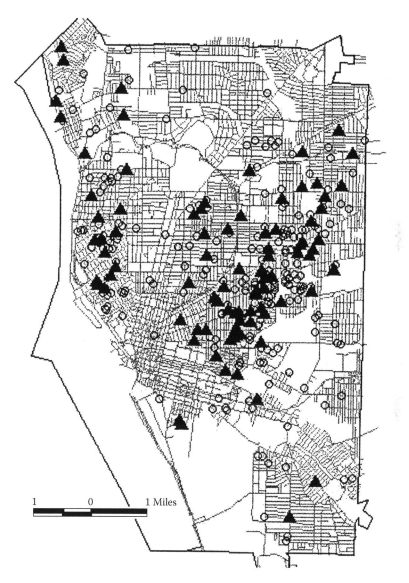

FIGURE 8.10
Arson locations before and after signal. Triangles indicate after.

The arson cluster alarm sounded in October 1996 implied that recent observations were locating nearer to previous arsons than expected. However, this could mean simply that the October 1996 arsons were located close to other arsons that were quite removed in a temporal sense (for example, perhaps the October 1996 arsons were located close to the location of January or February arsons). Although this type of monitoring will sometimes be of interest, it will also be of interest to monitor changes in the pattern of arsons that occur over specified windows of time to see if new observations are geographically close to those observed in the recent past.

Suppose, for example, that we wish to implement spatial surveillance using a temporal window containing only the previous 10 observations. Thus, we would be looking for an increase (or conceivably a decrease) in the degree of clustering, where the definition of clustering is based on the minimum distance from a newly observed point to any of the 10 previous observations. To implement this, we first find the minimum distance from a point, observed during the base period, to its nearest neighbor (where the set of nearest neighbors includes only the 10 most recent observations). We next compare that distance with the distance expected in a random pattern (again, obtained by taking the mean of a large number of minimum distances from randomly chosen points to sets of 10 successive points that have been observed during the base period). When surveillance begins, this process is continued, with the observed distance to the previous 10 observations being compared to the distance that would be expected if the base pattern did not change.

To illustrate, we define a subregion of the city of Buffalo where arson density appears the highest, and start surveillance in that subregion at observation 101 (after establishing a base pattern with the first 100 observations) with a moving window of 10 observations. Figure 8.11 indicates that there are two cluster signals over the remainder of the year. Figure 8.12 displays the locations of the observed arsons that

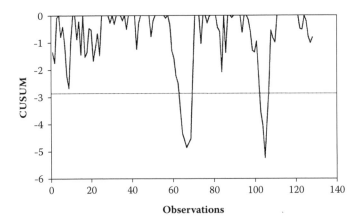

FIGURE 8.11
Cusum for the 1996 arson data with window of 10 observations (after 100 observations are used as a base period).

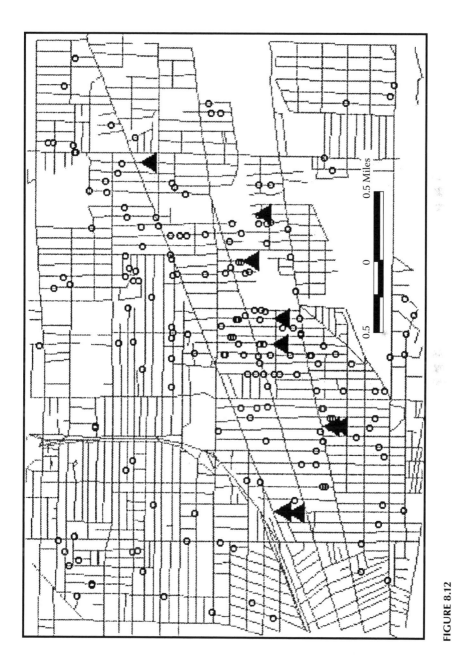

FIGURE 8.12
Arson locations leading to a cluster signal with window of 10 observations.

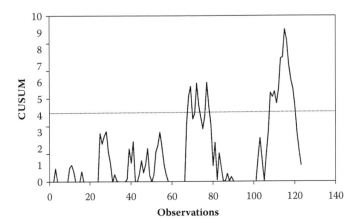

FIGURE 8.13
Cusum for monitoring dispersion tendency in the 1996 arson data with window of 10 observations.

occurred October 5–11, just prior to the second of these cluster signals. These arsons are located nearer to one another than would be
expected, given the usual distances observed between sequences of
10 arsons observed during the base period. Sensitivity to changes in the
base period was investigated by defining the first 50 observations and
the first 150 observations as the base period. In each of these alternatives, cluster signals were noted at the same times as those displayed in
Figure 8.11. Figure 8.13 displays the results of surveillance for *increases*
in the distances between neighbors. Two primary signals are sent—
the first after the 63rd observation following commencement of monitoring (on October 17th), and the second after the 107th observation
(on December 10th). Finally, Figure 8.14 portrays, using black triangles,
the observations leading up to the second of these two signals. These
arson locations tended to occur farther from their nearest neighbors
(where nearest neighbors are defined by the minimum distance to the
10 previous observations) than would be expected. Indeed, the black
triangles appear in this figure in relatively scattered locations within
the subregion. This change might have been either temporary or
more long lasting. The fact that the cusum returns to less than critical
values before observation 120 suggests that it was temporary. It is interesting that the signals for uniformity (large distances between arsons)
occurred immediately following the cluster signals. Perhaps the temporary change was the result of increased patrol in the area that had a
high density of arsons.

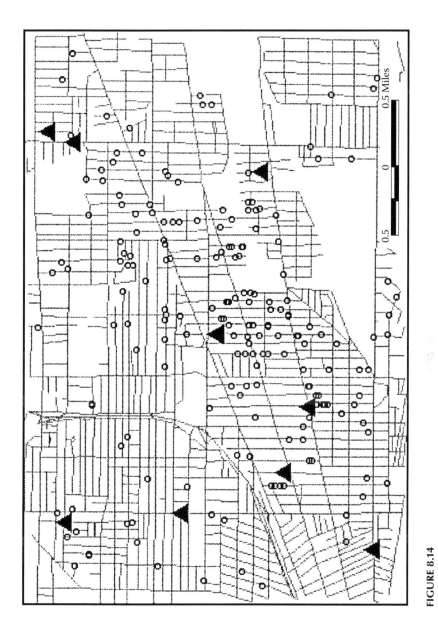

FIGURE 8.14

Arson locations leading to a dispersion signal with window of 10 observations (after 100 observations are used as a base period).

8.5   Summary and Discussion

In this section, we have described a procedure for monitoring changes in spatial patterns over time using a global spatial statistic and the cusum control chart scheme. The method results in the rapid detection of deviations from expected patterns. It may be used for various subregions of the study area, and it may be implemented using time windows of differing length to search for changes in spatial pattern that may occur at particular time scales. If X_t denotes any measure of spatial pattern at time t, we may use the following z-scores in a monitoring system that employs cumulative sums:

$$z = \frac{X_t - E[X_t \mid X_{t-1}]}{\sqrt{V[X_t \mid X_{t-1}]}} \tag{8.30}$$

An important issue concerns the choice of a base pattern. Ideally, the analyst should be able to specify with confidence some prior period of time that was in some sense stable with respect to the evolution of spatial pattern and that could serve as a basis for comparison. One would not likely want to choose an odd or unusual period of time as a base period; any subsequent changes that were detected might simply signal a return to normalcy.

It should be clear that this method does not give the analyst answers to the all-important question of *why* the change in pattern has occurred. It does provide, however, a way of signaling *when* a significant spatial change occurred. In the case of the arson example in Section 8.4.3, this should lead to both better short-term, strategic plans, and further hypotheses and investigations regarding the cause of change. In addition to signaling unexpected changes in patterns, it should also be of interest to detect changes in spatial patterns such as displacement that can be expected following targeted enforcement efforts. Although it would clearly have been interesting to investigate the possible causes of the changes in arson patterns in Buffalo (described in the previous subsections), this was unfortunately not possible.

The monitoring approach described here focuses upon changes in *geographical* patterns only. Thus, it does not signal either increases or decreases in the volume of criminal activity that may have taken place uniformly over the study region. This should not be viewed as a weakness of the method; the spatial monitoring method is designed to do exactly what its name implies, and should, of course, be combined with other appropriate analytic tools that achieve other objectives.

Although the illustrations used in this chapter have been applied to sets of past data, it should be emphasized that monitoring systems are designed primarily for handling new data as they become available. Although retrospective detection of pattern changes will certainly be of interest in many situations, it is the rapid detection of changes in *current* patterns that is of most interest.

Finally, we are not suggesting that investigators wait for changes in patterns to develop before they begin their investigations. For example, for some crime types, it will be important to follow any potential information, and a high "false-alarm" rate in the monitoring system may therefore be tolerable. The system described here provides just one of many important pieces of information and is designed to complement, rather than replace, other methods of analysis. Clearly, changes in spatial patterns may occur for many reasons. In some cases, investigators and analysts will possess "expert knowledge" that will be far more useful than a statistical analysis. However, there are many other cases where statistical monitoring of pattern changes should prove beneficial. Individuals are notoriously poor at detecting whether significant clusters exist on a map; there is a tendency to see clusters where none exist. It is, therefore, not a good idea to rely simply on visual interpretation. In addition, for situations with a high stream of data it could be easy to overlook changes in pattern.

9

Cusum Charts for Local Statistics and for the Simultaneous Monitoring of Many Regions

9.1 Monitoring around a Predefined Location

9.1.1 Introduction

In the previous chapter, cumulative sum (cusum) methods for detecting change (introduced in Chapter 7) were combined with global spatial statistics (introduced in Chapter 3) to monitor changes in spatial patterns. We begin this chapter by turning our focus to local spatial statistics (introduced in Chapter 4), where we describe how cusum methods may be used to monitor changes in the spatial pattern of events occurring around a predefined location of interest.

Raubertas (1989) was perhaps the first to suggest how cusum statistics may be used to monitor the possible evolution of clusters in a neighborhood around a point, with the objective of detecting any emergent spatial cluster as soon as possible after it has occurred.

Subsection 9.1.2 provides a summary of Raubertas' contribution. As his method is designed to monitor clustering for data aggregated to cells or subregions, it is also of interest to develop similar monitoring schemes for data where precise locations and precise times are available. In the subsections that follow, we discuss and compare two such tests, one similar to that of Raubertas and the other based on distances of cases from the focus. In subsection 9.1.6, the methods are applied to the clustering of cases of Burkitt's lymphoma. Subsection 9.1.7 discusses a method that is applicable when case-control data are available. The remainder of the chapter addresses the cases where separate charts are kept for each region and where many local statistics are monitored simultaneously.

9.1.2 Raubertas' Approach to Monitoring Local Statistics

Raubertas (1989) suggested the use of cusum techniques to monitor any changes in disease rate that occur in the geographical neighborhood of a region of interest. He suggests the monitoring of a weighted combination of Poisson random variables for region i, defined as

$$y_i = \sum_j w_{ij} x_j \qquad (9.1)$$

where x_j is the observed number of cases in region j, and w_{ij} is a measure of the connectedness of regions i and j. When the system is in control (i.e., when the region and its neighborhood experience no excess disease incidence), each x_j is taken as a Poisson random variable, with expectation equal to $\lambda_{0i} = n_i \pi_0$, where n_i is the population of region i, and π_0 is the baseline rate common to all regions. When the rate has increased to a level where detection is desirable (say, π_1), $\lambda_{1i} = n_i \pi_1$ is the mean number of cases expected. The cusum for region i is

$$S_{i,t} = \max(0, S_{i,t-1} + y_{i,t} - k) \qquad (9.2)$$

As noted in Chapter 7, subsection 7.5.1, Lucas (1985) provides values of the critical parameters h and k for various combinations of λ_0 and λ_1 when the desired ARL_0 is equal to 500 and the desired ARL_1 is equal to 7. For other parameter values, one could use the method of Brook and Evans (1972) to find the average run length. Their method relies upon a Markov chain transition probability matrix, where the elements are functions of the probability that the cusum changes from one value to another. Critical values of h for other desired ARLs could also be found via Monte Carlo simulation.

9.1.3 Monitoring a Single Local Statistic: Autocorrelated Regional Variables

Suppose there is no spatial autocorrelation in the regional values being monitored and we suspect that, when change occurs, it will occur in the form of increases in a subset of regions comprising a neighborhood rather than in a single region. There are at least two ways forward if our objective is to detect this increase quickly:

1. Keep a single chart for the variable consisting of a weighted sum of the regional values (similar to the suggestion of Raubertas).
2. Use the approach of Healy (1987) described in the following text, which is optimal for quick detection of change in a single, hypothesized direction.

Although these approaches should give identical results under the conditions specified, Healy's approach is more general because it can also handle the situation where the underlying variables are correlated. Specifically, when the variance–covariance matrix associated with the regional values is designated Σ, the following cusum based on vectors of regional observations (x_t) is optimal for detecting a change in mean from μ_G to μ_B, where these latter quantities are vectors of regional values for the good, in-control, and bad, out-of-control, means, respectively:

$$S_t = \max(0, S_{t-1} + \mathbf{a}'(\mathbf{x}_t - \boldsymbol{\mu}_G) - 0.5D) \tag{9.3}$$

where

$$\mathbf{a}' = \frac{(\boldsymbol{\mu}_B - \boldsymbol{\mu}_G)' \Sigma^{-1}}{\{(\boldsymbol{\mu}_B - \boldsymbol{\mu}_G)' \Sigma^{-1}(\boldsymbol{\mu}_B - \boldsymbol{\mu}_G)\}^{1/2}} \tag{9.4}$$

and

$$D = \sqrt{(\boldsymbol{\mu}_B - \boldsymbol{\mu}_G)' \Sigma^{-1}(\boldsymbol{\mu}_B - \boldsymbol{\mu}_G)} \tag{9.5}$$

9.1.4 An Approach Based on Score Statistics

An alternative that is similar to Raubertas' approach is to make use of the score statistic (see Chapter 4, Section 4.3):

$$U_i = \sum_{j=1}^{m} w_{ij}(C_j - np_j) \tag{9.6}$$

where C_j is the observed number of cases in region j, n is the total number of cases, and p_j is the probability that an individual case falls within region j under the null hypothesis of no clustering around the focus (designated as region i). The quantity np_j is therefore the expected number of cases that fall into region j. Furthermore, w_{ij} is a weight attached to the importance of region j in relation to the focus region i; generally, this will decrease as one goes out to more distant regions from i. Note that Equation 9.6 is equivalent to Equation 4.4 (Chapter 4).

Under the null hypothesis of no clustering around the focus, if the number of cases is sufficiently large, U_i has a normal distribution (e.g., see Tango 1995) with mean zero and variance given by Equation 4.5.

Because we will use this statistic on successive cases, we will use $n = 1$ and, for each new case, define C_j to be equal to one if the case is in region j, and zero otherwise. We then cumulate the quantity

$$S_{i,t} = S_{i,t-1} + \frac{U_i}{\sqrt{V[U_i]}} - k \tag{9.7}$$

where $S_{i,t}$ is the cusum for the focus region i at time t and $S_{i,0} = 0$. As with other examples, batching of observations may be necessary to confer normality.

This approach is similar to that of Raubertas, where the difference is that a normal approximation is being used.

9.1.5 Spatial Surveillance around Foci:
A Generalized Score Statistic, Tango's C_F

As noted in Chapter 4, Section 4.4, Tango (1995) suggested a modified and generalized score statistic to test the null hypothesis of no clustering against the alternative of clustering around prespecified "foci." His statistic is denoted by C_F and is given by Equation 4.7, with variance given by Equation 4.8. Under the null hypothesis, $C_F/\sqrt{\text{var}(C_F)}$ has, asymptotically, a standard normal distribution as the number of cases n gets large.

The cusum used for surveillance around the focus or foci is now based upon the quantities

$$Z_i = \frac{C_{F,i} - E[C_{F,i} \mid C_{F,i-1}]}{\sigma_{C_{F,i} \mid C_{F,i-1}}} \tag{9.8}$$

where $C_{F,i}$ is Tango's C_F statistic as calculated after the first i observations. Also, the conditional expectation is

$$E[C_{F,i} \mid C_{F,i-1}] = \mathbf{p}'\mathbf{s} \tag{9.9}$$

where $\mathbf{s}$ is a $m \times 1$ vector containing as element k

$$s_k = \{\mathbf{W}'\mathbf{A}(\mathbf{r}_{i-1}(k) - \mathbf{p})\} \tag{9.10}$$

and where $\mathbf{r}_{i-1}(k)$ is the proportion of cases in each region after i observations, given $\mathbf{r}_{i-1}$ (which denotes the proportion of cases in each region after $i - 1$ observations), and given that case i is located in region k. The conditional variance is

$$\sigma^2_{C_{F,i} \mid C_{F,i-1}} = \mathbf{p}'(\text{diag } \mathbf{s}'\mathbf{s}) - (\mathbf{p}'\mathbf{s})^2 \tag{9.11}$$

If necessary, we cumulate the means of batches of Z_i's.

We next used the four scenarios described in Table 8.4 for monitoring general clustering to assess the performance of this surveillance plan for focused clustering. We simulated 1000 repetitions of each scenario. Experimentation with different batch sizes (b) revealed that convergence to normality was slower than for general clustering. With $b = 4$, a significant amount of positive or negative skewness was often present in the $Z_{(b)}$ values under the null hypothesis. Positive skewness reduces ARLs, whereas negative skewness has the effect of making the test conservative in the sense that ARLs are higher than they are under normality. We used $b = 10$ in the results that follow because convergence to normality was usually achieved for that batch size. No serial dependence was detected among 250 observations of $Z_{(10)}$ simulated under the null hypothesis. It should also be mentioned that the focus under each scenario was selected to match the point around which clustering was actually simulated (although it is unlikely to be known in advance of the analysis).

The results are shown in Table 9.1. A comparison with Table 8.4 shows that focused surveillance is more likely than general surveillance to find evidence of clustering within the first 100 observations, when it, in fact,

TABLE 9.1

Results of Cumulative Sum Surveillance with Tango's C_F Statistic on Simulated Data

	Probability of Rejecting				
	H_0 after $n = 100$	Observations		Median	ARL
Focused Tests	Tango (C_F)	Cusum	ARL	Run Length	RL < 100
Simulated Clustering:					
None	.056	.060	—	—	—
Around Point A	.847	.792	73.5	60	51.2
Around Point B	.885	.828	69.5	60	52.0
Around Point C	.954	.893	59.2	50	49.6
Around Point C ($b = 4$)	—	.908	51.6	44	—

exists. Average run lengths are also lower in Table 9.1 than in Table 8.4, as expected. Surveillance for increased incidence around a point leads to quicker detection when there is an elevated actual relative risk there, in comparison with surveillance for general clustering. The power of Tango's retrospective test is again slightly higher, but the surveillance scheme has the advantage of its implementation as observations become available, and is designed for rapid detection of changes.

Finally, note from the last two rows of this panel that the amount of skewness present in the distribution of Z in this instance does not appear to affect the ARL severely.

9.1.6 A Distance-Based Method

When precise locations and times of occurrence are known, alternative cusum schemes that make use of this information may be constructed. With precise data on time of occurrence, there is no need to construct artificial time periods. In addition, when data on precise locations are known, there is no real need to deal with artificially constructed regions.

Suppose we had information on the precise location of new cases as they occurred (in contrast with the situation where we only know the region in which the new case falls, as in subsections 9.1.2 through 9.1.5). If the study region is circular around the focus, and we wish to monitor the successive locations of points for deviations from spatial randomness, we may make use of the distribution of distances from the center of a circle to random locations within the circle. Specifically, the expected distance from the center of a circle of radius R to a randomly placed point is $2R/3$, and the variance of such distances is $R^2/18$. To monitor for increases in clustering near the center (and hence for distances that are *less* than expected), we may use the cusum

$$S_t = \min(0, S_{t-1} + z_t + k) \tag{9.12}$$

where

$$z_t = \frac{d_t - \frac{2R}{3}}{\sqrt{1/18R}} \tag{9.13}$$

and where d_t is the observed distance to the point observed at time t (where the index t is used to indicate the observation number).

One question concerns the underlying assumption of normality. The distance distribution is *not* normally distributed, and again it is possible to aggregate observations into batches of b observations. Then, the mean of the b observations can be used with

$$S_t = S_{t-1} + \bar{z}_{(b)} + \frac{k}{\sqrt{b}} \tag{9.14}$$

with signals occurring when

$$S_t < \frac{-h}{\sqrt{b}} \tag{9.15}$$

Because of spatial variations in the distribution of population at risk, it is unlikely that we will wish to monitor deviations from complete spatial randomness. Rather, we might wish to monitor for deviations relative to the distribution of the at-risk population. In this case, we would replace the values of $2R/3$ and $R^2/18$ mentioned previously with the mean and variance of the distance to the at-risk population within the circular study region. Alternatively, we might wish to monitor for deviations from some base pattern of cases witnessed during a "base" period; here, we would replace the mentioned mean and variance with the mean and variance of the distances observed during the base period.

9.1.6.1   Application to Data on Burkitt's Lymphoma

To examine spatial patterns around a particular location using the data on Burkitt's lymphoma (see subsection 8.4.1.2), Arua hospital in the region of Ayivu was chosen as the focus because it is one of the two major hospitals in the region. Such an examination is interesting and relevant because as the authors point out, some of the spatial variation in cases may be due to the improving likelihood of detecting cases near hospitals.

The base period consisted of the first 80 cases (which occurred during the period January 1, 1961 to February 28, 1969). The mean distance from the hospital in Arua to cases during this period was 39.2 km, and the standard deviation of the distances was 23.1 km. Monitoring commenced on March 1, 1969. Cases were batched into groups of three, and a value of $h = 4/\sqrt{3}$ was chosen. To determine the value of ARL_0 implied by this choice of h, the null hypothesis of no clustering relative to the base period was first simulated. The 80 base-period cases were grouped into eight rings corresponding to radii of 10, 20, ..., 80 km around the focus. Then, using the relative frequency of base-period cases in these rings, hypothetical new cases were chosen and their distances from the focus (denoted d) subjected to a cusum using $z = (d - 39.2)/23.1$. This corresponded to an ARL_0 of approximately 1500. Figure 9.1 displays the cusum. The first clustering signal is sent after observation 146, which occurs in March 1973. The mean distance to cases for the eight observations that immediately preceded this signal was 23.0 km, significantly less than the mean of 39.2 km observed during the base period. It is interesting to note that the cusum remains below the critical value of h for the remainder of the study period. The average distance to observed cases during this postalarm period was 31.1 km, still substantially less than that observed during the base period.

This cluster was also found in the original, retrospective analysis of the data. In particular, a cluster was found in this region for data aggregated into the time period 1972–1973. Had a monitoring scheme been in place as the data were collected, the cluster would have been detected in March 1973— much sooner than the time of detection under the retrospective analysis.

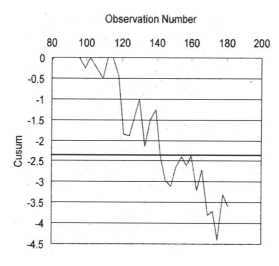

FIGURE 9.1
Cumulative sum associated with monitoring of clustering around Arua.

9.1.7 Surveillance around Prespecified Locations Using Case–Control Data

9.1.7.1 Introduction

Up to this point in the chapter, we have not yet looked at how we might go about monitoring around a prespecified location when the data come in the form of cases and controls. In this section, we assume that the data on the locations of cases and controls is arranged in temporal order (for example, data might consist of the date of diagnosis for both case and control diseases). We then illustrate how conventional modeling approaches may be adapted to use the dataset observation by observation, to detect as quickly as possible a change from one set of model parameters to another.

9.1.7.2 Prospective Monitoring around a Source, Using Case–Control Data

These methods are optimal for detecting step changes in the model parameters. They are based on the score statistic z_t, which represents the difference in log-likelihoods of the observations before and after the change, and in turn is derived from the observations x_t at time t:

$$z_t = \ln\left[\frac{f(x_t \mid \theta^{(1)})}{f(x_t \mid \theta^{(0)})}\right] = \ln f(x_t \mid \theta^{(1)}) - \ln f(x_t \mid \theta^{(0)}) \qquad (9.16)$$

where f designates the likelihood function, and $\theta^{(0)}$ and $\theta^{(1)}$ refer to the vector of parameters before and after the change, respectively.

These scores are then used to formulate the cumulative sum S_t:

$$S_t = \max(0, S_{t-1} + z_t) \qquad (9.17)$$

A change from $\theta^{(0)}$ to $\theta^{(1)}$ is detected when the cumulative sum S_t exceeds the predefined threshold h.

Suppose that, for monitoring case–control data, we adopt Diggle's function (Chapter 4, Equation 4.14), and set $\theta^{(0)} = \{\rho, \theta_1^{(0)}, \theta_2^{(0)}\} = \{\rho, 0, 0\}$, implying no raised incidence around the prespecified source prior to the change (recall that ρ represents the ratio of cases to controls the parameter θ_1 estimates the excess risk at the source, and the parameter θ_2 represents exponential decline in risk as one travels away from the source). Let $\theta^{(1)} = \{\rho, \theta_1^{(1)}, \theta_2^{(1)}\}$ be the parameters after the change. The likelihood of observing a control at a distance r from the source under the new regime is, using Equation 4.16,

$$L(\theta^{(1)}) = -\ln\left\{1 + \rho\left(1 + \theta_1^{(1)} e^{-\theta_2^{(1)} r^2}\right)\right\} \qquad (9.18)$$

The likelihood of observing a control under the initial regime implies, using Equation 4.17 and the adoption of $\theta^{(0)} = \{\rho, 0, 0\}$,

$$L(\theta^{(0)}) = \ln(\rho) - \ln(1 + \rho) \qquad (9.19)$$

Similarly, observation of a case under the new regime using Equation 4.16 has the likelihood

$$L(\theta^{(1)}) = \ln \rho + \ln\left(1 + \theta_1^{(1)} e^{-\theta_2^{(1)} r^2}\right) - \ln\left\{1 + \rho\left(1 + \theta_1^{(1)} e^{-\theta_2^{(1)} r^2}\right)\right\} \qquad (9.20)$$

and observation of a case under the initial regime, using Equation 4.17 and the adoption of $\theta^{(0)} = \{\rho, 0, 0\}$, leads to the following likelihood:

$$L(\theta^{(0)}) = \ln(\rho) - \ln(1 + \rho) \qquad (9.21)$$

These four expressions in Equations 9.18 through 9.21 can then be used to form the likelihood ratio and hence the score statistic for new observations as they become available.

9.1.7.3   *Illustration*

To illustrate the use of monitoring in this context, we adopt a circular study area of radius one surrounding the hypothetical putative source. Observations were assigned case status with probability one-half and control status with probability one-half, and consequently we take the value of ρ under both the null and alternative hypotheses to be equal to one. Population is assumed to be distributed uniformly, and therefore cases and controls are generated under the null hypothesis by choosing locations at random within the study area.

To simulate the distances that cases and controls lie from the source, the following approach, based on the cumulative distribution function of distances, was used:

1. Take a randomly chosen number u from a uniform distribution on the interval $(0, 1)$.
2. Set it equal to the cumulative distribution function associated with the distribution of distances.
3. Solve for the random variable representing distance from source.

Thus, the uniform random number u is set equal to the probability that an observation lies within a distance r (less than or equal to R, which is the radius of the study area set to be one in this illustration) of the source. For both cases and controls under the null hypothesis,

$$u = F(r) = \frac{\int_0^r 2\pi x \, dx}{\int_0^R 2\pi x \, dx} = \frac{r^2/2}{1/2} = r^2 \tag{9.22}$$

Hence, we can simply take as the simulated distance the square root of a random number chosen from a uniform distribution on the interval $(0,1)$ because $r = \sqrt{u}$. The result represents the distance from the source at the center of the study area (the direction, and hence, precise location is not important because we are only interested in the distance from the source).

Under the alternative hypothesis, controls are generated in the same way, but cases are now chosen in such a way that they are more likely to occur near the putative source. Using the function suggested by Diggle (Chapter 4, Equation 4.14),

$$u = F(r) = \frac{2\pi \int_0^r (1 + \theta_1 e^{-\theta_2 x^2}) x \, dx}{2\pi \int_0^{R=1} (1 + \theta_1 e^{-\theta_2 x^2}) x \, dx} \tag{9.23}$$

which leads to

$$u = F(r) = \frac{\theta_1 + \theta_2 r^2 - \theta_1 e^{-\theta_2 r^2}}{\theta_1 + \theta_2 R^2 - \theta_1 e^{-\theta_2 R^2}} = \frac{\theta_1 + \theta_2 r^2 - \theta_1 e^{-\theta_2 r^2}}{\theta_1 + \theta_2 - \theta_1 e^{-\theta_2}} \tag{9.24}$$

where the latter term on the right-hand side results from our choice of $R = 1$. For any choice of the random number u on the interval $(0,1)$, we wish to solve for r. Because it is not possible to solve for r directly as a function of u, the solution for r is achieved by numerical root-finding methods (see appendix to this chapter).

Alternatively, it is possible to find an approximation for r in terms of u. First rearrange Equation 9.24 as

$$u(\theta_1 + \theta_2 - \theta_1 e^{-\theta_2}) - \theta_1 = \theta_2 r^2 - \theta_1 e^{-\theta_2 r^2} \tag{9.25}$$

Designating the left-hand side by y and solving this for r^2 in Maple 9.5 yields

$$r^2 = \frac{y + LambertW(0, \theta_1 e^{-y})}{\theta_2} \tag{9.26}$$

For a given argument x, the LambertW function (Corless et al. 1996) returns the (possibly multiple) values $W(x)$ satisfying $W(x)\,e^{W(x)} = x$. For example, when $x = -0.1$, $W(x) = -3.577$, and $W(x) = -0.1118$ represent solutions. In Equation 9.26, the first argument of "0" refers to a particular branch (in fact, the main branch) of the multivalued function, and the second term is the argument of the function.

There are alternative approaches to the numerical evaluation of the LambertW function (Chapeau-Blondeau and Monir 2002). There are also various series expansions for the LambertW function (Corless et al. 1997); an evaluation of these reveals that different series are most accurate across different ranges of the argument. In particular, for various values of the argument z, LambertW$(0, z)$ may be approximated via the following series expansions:

$$LambertW(0, z) \approx z - z^2 + z^3/3 - \qquad z \le 0.3$$

$$\approx \frac{2z}{z+e} + \frac{z(z-e)^2}{2(z+e)^3} + \qquad 0.3 < z \le 4 \qquad (9.27)$$

$$\approx v + \frac{vp}{1+v} + \frac{vp^2}{2(1+v)^3} + \qquad z > 4$$

where $v = \ln z$, and $p = -\ln(\ln z)$. When $z < 0.3$, combining Equation 9.26 and the first few terms of Equation 9.27 yields

$$r^2 \approx \frac{y + \theta_1 e^{-y} - (\theta_1 e^{-y})^2}{\theta_2} \approx \frac{y + \theta_1 e^{-y}}{\theta_2} \qquad (9.28)$$

Using just the first term on the RHS of Equation 9.27 for the middle range of z yields

$$r^2 \approx \frac{y + 2q/(q+1)}{\theta_2}$$

where $q = \theta_1 e^{-y-1}$. Finally, when z is large, using the first two terms of the last approximation in Equation 9.27 leads to

$$r^2 \approx \frac{y + w - (w \ln w)/(1+w)}{\theta_2}$$

where $w = \ln \theta_1 - y$.

For example, suppose we wish to simulate for the scenario where $\theta_1 = 2$, $\theta_2 = 4$. Suppose the random number is $u = 0.5$. Then, $y = u(\theta_1 + \theta_2 - \theta_1 e^{-\theta_2}) - \theta_1 = 0.982$ using Equation 9.25, and $r^2 = .771$ or $.768$, using the two- and three-term numerators of Equation 9.27, respectively.

After distances have been simulated, the scores (z_t) are computed. Observation of a control leads to a score found by subtracting Equation 9.19 from Equation 9.18:

$$z_t = -\ln\left(1+\left(1+\theta_1^{(1)}e^{-\theta_2^{(1)}r_t^2}\right)\right)-\ln(1)+\ln(2)$$

$$= -\ln\left(2+\theta_1^{(1)}e^{-\theta_2^{(1)}r_t^2}\right)+\ln(2) \tag{9.29}$$

Observation of a case leads to a score found by subtracting Equation 9.21 from Equation 9.20:

$$z_t = \ln(1)+\ln(1+e^{-\theta_2^{(1)}r_t^2})-\ln(1+1+e^{-\theta_2^{(1)}r_t^2})-\ln(1)+\ln(2)$$

$$= \ln(1+e^{-\theta_2^{(1)}r_t^2})-\ln(2+e^{-\theta_2^{(1)}r_t^2})+\ln(2) \tag{9.30}$$

These scores are then used in the cusum.

We now illustrate the procedure and results for the choices $\theta^{(0)} = \{1,0,0\}$ and $\theta^{(1)} = \{1,2,4\}$. The null hypothesis is simulated by first assigning case/control status (using ½ as the probability an observation is a case) and then choosing distances $r = \sqrt{u}$ for both cases and controls. These distances are then used to determine scores (Equations 9.29 and 9.30), and the cusum is run until it reaches a threshold h. Suppose that we desire an average run length (ARL) of 250 observations between false alarms—that is, declarations of change when in fact none has occurred. Experimentation with different threshold values revealed that a value of $h = 2.0$ is consistent with an in-control ARL of approximately 250. Next, the alternative hypothesis was simulated by using distances determined from Equation 9.24 using parameters equal to those chosen for $\theta^{(1)} = \{1,2,4\}$. The ARL under this alternative hypothesis was approximately 103; this is the number of observations that would be required on average to detect the change in risk.

For more pronounced increases in risk near the source, detection occurs more quickly, as would be expected. For example, with $\theta^{(1)} = \{1,6,5\}$, a threshold of $h = 2.7$ leads to an ARL of approximately 250 (making this instance comparable with the previous one); when the alternative hypothesis is simulated using the chosen values of $\theta^{(1)}$, the average time to detection declines to approximately 60 observations.

It will, of course, not usually be possible to specify correctly the magnitude of the shift. For example, suppose that a shift to $\theta^{(1)} = \{1,1,3\}$ is hypothesized, but the actual shift is to $\{1,2,4\}$. A threshold of $h = 1.59$ leads to an ARL of 250 under the null hypothesis, and now the average time to detect a change is approximately 110—slightly longer than the 103 found earlier when a correct estimate of the shift is adopted (using $\theta^{(1)} = \{1,2,4\}$). Similarly, if a shift to $\theta^{(1)} = \{1,3,5\}$ is hypothesized, a threshold of $h = 2.14$ leads to an ARL of 250 under the null hypothesis of no change, and the average time to detect the shift to $\{1, 2, 4\}$ is approximately 107. At least in these examples, misestimation of the magnitude of the shift does not affect significantly the time to detection.

We have simplified the problem by assuming that ρ does not change. This is analogous to cusum procedures that are optimized for detection of changes in the mean, where it is assumed that the variance does not change. Any change in ρ over time will, of course, render the cusum scheme less effective. We have also focused here solely on the spatial aspects of the process. Although we have been interested here in the unfolding of spatial patterns over time, we have essentially ignored the temporal aspect of the process; that is, our interest has been purely in whether or not the risk was increased around the putative source while assuming that the overall intensity of the disease was constant over time. In any analysis of disease, however, attention must, of course, also be given to how the intensity of the disease varies over time.

9.2 Spatial Surveillance: Separate Charts for Each Region

To this point, we have covered the monitoring of both global and local statistics. The third category of statistical tests in the Besag–Newell classification concerns the detection of clustering, which includes the scan-type tests of Chapter 5, where the objective is to assess many local statistics simultaneously. Similarly, an important issue in spatial surveillance and monitoring concerns the common desire to monitor more than one location at once. One straightforward way to monitor variables observed for a set of regional subunits over time is to maintain a separate chart for each subunit. An immediate issue that arises in the context of monitoring across a set of regional subunits is how to properly account for the multiple testing across spatial units.

Let us first illustrate this for the case of the Shewhart charts. For an individual chart, the use of a threshold of $z = 1.645$ would lead to a false alarm once every $1/0.05 = 20$ time periods (because 0.05 is the tail probability associated with this z-value, and hence the probability of signaling on any individual observation). Now suppose that, each time period, we have observations in the form of z-scores for each of $m = 100$ regions. Further, suppose that we wish to retain the average false-alarm rate of one every 20 periods—that is, we would like to have no false signals anywhere in the system, but we are willing to tolerate, on average, one such alarm every 20 periods. Continued use of $z = 1.645$ as a threshold would lead to alarms at a rate that is much more frequent than desired; as there are so many regions, it becomes much more likely that *at least one of the regions* will signal. One solution is to use a Bonferroni adjustment. Rather than using .05

as the tail probability, we use $.05/m$. In this case, using a tail probability of $.05/100 = .0005$ leads to a threshold of $z = 3.28$. This, in turn, leads to a desired false alarm, on average, once every $1/0.05 = 20$ time periods.

If the desired frequency of false alarms differs from this case, an adjustment to this can easily be made. In general, if the desired time between false alarms is denoted ARL_0 then, for a single chart, $\alpha = 1/ARL_0$. For example, suppose that one wishes to have an average of 50 time periods between false alarms. Then, it would be appropriate to use $\alpha = 1/50 = .02$ in this case. For m charts, one uses the value of z that leaves $(1/ARL)/m$ in the tail of the normal distribution. For the value of $m = 100$ in the present example, this corresponds to a critical z-value of $\Phi^{-1}(.02/100) = 3.54$ to be used as the threshold on each of the $m = 100$ charts.

If cusum control charts are kept simultaneously for many regions, the average run length between false alarms (where a signal is received from any one of the many charts) will be less than that implied by the usual threshold derived for a single chart (where the threshold in turn is based upon the desired ARL). More precisely, the average run length between false alarms for the combined set of m charts (one for each region), ARL_0^*, will be (Raubertas 1989):

$$ARL_0^* = \frac{1}{1-(1-1/ARL_0)^m} \tag{9.31}$$

This is based upon the fact that the time between false alarms has an exponential distribution (Page 1954), and hence, the probability that any single observation on a single regional chart leads to a false alarm is $1/ARL_0$. The probability of no alarm following a single observation in a region is therefore $1 - 1/ARL_0$, and the probability that none of the m regions signals during a given time period is $(1 - 1/ARL_0)^m$. Thus the probability that at least one of the m charts signals is $1 - (1 - 1/ARL_0)^m$, and the inverse of that is the average time until a signal. Alternatively stated, the ARL_0 to use on each chart is given by

$$ARL_0 = \left[1 - \left(1 - \frac{1}{ARL_0^*} \right)^{1/m} \right]^{-1} \tag{9.32}$$

where, again, ARL_0^* is the desired time between false alarms from any of the m regions. This is analogous to the Sidak adjustment that was discussed in Chapter 5. A computationally simpler way to account for the simultaneous monitoring of the m charts is to use a Bonferroni-type adjustment; instead of using Equation 9.32 to determine the threshold for each chart, the quantity

$$ARL_0 = m\,ARL_0^* \tag{9.33}$$

is used. Thus, if there are $m = 10$ regional units and a desired time between false alarms of $ARL_0^* = 100$, the threshold for each chart is found using $ARL_0 = 10 \, (100) = 1000$, and then employing Equation 7.4. Equations 9.32 and 9.33 give very similar results in practice.

9.2.1   Illustration

To illustrate how the cusum methodology may be implemented in a multiregional context, and to illustrate some of the issues that arise, data were simulated for a nine-region spatial system constructed by assuming a three-by-three structure of square regions in a square study area. This is a multiregional extension of the example given in Chapter 7, subsection 7.3.1 (where the data for region 1 only are used). Simulated data are in the form of standardized z-scores. The regions were numbered from one to nine, beginning in the upper-left hand corner and proceeding row by row, with the lower right subregion designated as subregion nine (a map of this hypothetical spatial system is shown in Figure 9.2). In Table 9.2, the simulated z-scores are depicted for each region for each time period. Each column represents a region, and each row represents a time period. The data in Table 9.2 were developed by first choosing random variates from a standard normal distribution for the first 15 time periods for each of the nine regions. Beginning in period 16, each region's mean value increased; the mean increased by 0.2 in regions 1, 3, 7, and 9; by 0.3 in regions 2, 4, 6, and 8, and by 0.75 in subregion 5. This corresponds to an increase that is centered on subregion 5, and dampens as one goes outward from there.

Now suppose that each region maintains its own cusum. If we use $k = 0.5$, and assume a desired ARL_0 of 100, Equation 7.4 implies a threshold of $h = 2.84$ that will be used in each region.

Region 1	Region 2	Region 3
(0.2)	(0.3)	(0.2)
Region 4	Region 5	Region 6
(0.3)	(0.75)	(0.3)
Region 7	Region 8	Region 9
(0.2)	(0.3)	(0.2)

FIGURE 9.2
Hypothetical nine-region system for an illustration of multiregional cusum method; values in parentheses indicate increases in the mean from time period 16.

TABLE 9.2

Hypothetical z-Scores for Nine Hypothetical Regions, for 30 Time Periods

Time Period	Region 1	Region 2	Region 3	Region 4	Region 5	Region 6	Region 7	Region 8	Region 9
1	−0.8	−2.04	−0.69	−1.47	1.05	−0.05	1.34	−0.52	1.58
2	0.3	0.42	−0.44	0.96	0.07	−0.2	−1.2	−0.44	0.15
3	0.31	0.11	−0.29	−0.01	−1.28	1.47	0.4	0.23	0.84
4	1.76	−1.65	−0.99	0.78	−0.85	0.54	0.82	−0.39	0.95
5	−0.75	1.48	−0.34	0.41	−1.33	−1.12	0.64	−1.41	0.65
6	0.38	0.05	−0.41	−0.07	−0.05	1.02	−1.6	0.23	0.91
7	0.57	−0.78	−1.4	0.25	−1.97	1.02	0.7	0.19	1.9
8	−0.79	−1.04	−1.34	−0.32	−0.13	0.48	−0.64	1.18	1.78
9	−1.04	1.14	−3.37	0.47	−0.88	−0.18	−1.83	1.11	0.55
10	1.14	−1.17	−1	0.61	1.62	0.48	−0.1	0.7	−0.03
11	1.4	0.03	0.94	−1.8	0.33	−0.91	−1.13	1.25	−0.66
12	0.35	−0.28	0.55	−0.18	0.37	−0.95	−0.22	−2.13	−2.56
13	1.38	−1.12	0	−1.11	0.37	−0.63	0.12	0.44	1.18
14	−0.01	0.83	−0.98	−0.55	0.17	−0.97	0.62	−0.24	0.74
15	1.56	1.67	1.73	2.89	−0.03	0.9	−0.39	−0.64	0.51
16	−0.57	−0.85	0.82	0.52	1.25	−1.03	0.97	−1.54	0.12
17	1.98	0.4	−0.01	1.45	0.13	0.02	0.36	0.93	0.92
18	1.77	−0.66	−0.12	0.33	0.27	1.1	−1.58	1.08	0.84
19	−0.89	0.07	0.09	0.55	0	−1.13	−0.02	1.51	−2.3
20	−0.81	1.4	0.61	1.69	−0.35	0.38	−1.02	1.18	0.76
21	−0.52	2.05	−0.11	1.59	1.21	−1.12	−0.57	−0.78	0.35
22	1.18	1.75	−0.37	−0.09	−0.74	−0.08	0.87	0.76	0
23	0.55	−0.01	0.41	1.65	1.87	1.15	−0.13	−0.59	−0.05
24	0.63	−1.29	0.87	1.65	2.26	−0.05	−0.68	−2.12	0.1
25	−0.2	1.12	1.94	−1.74	1.65	0.44	1.15	−0.3	−0.38
26	−0.44	−0.52	1.22	−0.56	0.3	0.73	2.4	−0.21	−0.25
27	0.65	−0.8	0.1	0.5	1.65	−0.36	−0.7	−1.47	−0.77
28	−1.99	1.92	−1.9	0.54	0.94	0.68	1.17	0.26	−1.71
29	1.28	0.77	0.44	−0.35	1.7	−0.29	−1.88	−1.07	−0.4
30	0.06	0.38	−0.93	−0.33	0.47	−0.37	−0.14	0.73	−1.74

Maintaining cusums for each region leads to signals where the threshold h is exceeded in the following instances:

Region 1: Periods 17–19
Region 2: Periods 22–23
Region 4: Periods 17–29
Region 5: Periods 24–30
Region 9: Periods 7–11

For illustration, the computation of the cusum values for regions 1 and 9 is given in Table 9.3.

TABLE 9.3

Cusum Computation for Subregions 1 and 9

	Region 1		Region 9	
Time t	$S_{t-1} + z_t - k$	$S_t = \max$ $(0, S_{t-1} + Z_t - k)$	$S_{t-1} + z_t - k$	$S_t = \max$ $(0, S_{t-1} + Z_t - k)$
1	−1.3	0	1.08	1.08
2	−0.2	0	0.73	0.73
3	−0.19	0	1.07	1.07
4	1.26	1.26	1.52	1.52
5	0.01	0.01	1.67	1.67
6	−0.11	0	2.08	2.08
7	0.07	0.07	3.48	*3.48*
8	−1.22	0	4.76	*4.76*
9	−1.54	0	4.81	*4.81*
10	0.64	0.64	4.28	*4.28*
11	1.54	1.54	3.12	*3.12*
12	1.39	1.39	0.06	0.06
13	2.27	2.27	0.74	0.74
14	1.76	1.76	0.98	0.98
15	2.82	2.82	0.99	0.99
16	1.75	1.75	0.61	0.61
17	3.23	3.23	1.03	1.03
18	4.5	*4.5*	1.37	1.37
19	3.11	*3.11*	−1.43	0
20	1.8	1.8	0.26	0.26
21	0.78	0.78	0.11	0.11
22	1.46	1.46	−0.39	0
23	1.51	1.51	−0.55	0
24	1.64	1.64	−0.4	0
25	0.94	0.94	−0.88	0
26	0	0	−0.75	0
27	0.15	0.15	−1.27	0
28	−2.34	0	−2.21	0
29	0.78	0.78	−0.9	0
30	0.34	0.34	−2.24	0

Note: Italic characters indicate cusum values greater than the threshold value (2.84).

It should also be pointed out that one of the regions (region 9) signals even before the change occurs at period 16, and the region's cusum value quickly increases at the very beginning of the simulation in Table 9.3. This is clearly a false alarm. Three other regions (1, 2, and 4) exceed the threshold, but only for a temporary period. By time period 30, an increased mean is indicated only in region 5.

These nine separate surveillance systems might be suitable for each of nine individual, regional health departments. However, there are a number of important aspects of surveillance pertaining to the spatial and hierarchical structure of the study area that merit further discussion. For example:

1. A state health official desiring an ARL_0 of 100 (that is, an average time of 100 time periods before witnessing an alarm in *any* of the nine regions) would have to set the threshold higher than the value of 2.84 found earlier; otherwise, alarms would occur too frequently.

2. Regional officials could conceivably miss a change that is spread across several regions. Note in the foregoing example that small changes have occurred in each region but the regional alarms are not necessarily persistent or timely. In general, the magnitude of the change might be relatively small in any particular region, but if many such small changes across clusters of counties are viewed in their totality, the change may become more apparent and detectable. This is discussed further in Section 9.3.

For the example, with $ARL_0^* = 100$ and $m = 9$, the average run length is found to be $ARL_0 = 895.99$, using Equation 9.32. This leads to $h = 4.95$, using Equation 8.4.

For the data in Table 9.2, only the cusums for regions 4 and 5 attain this threshold:

Region 4: Periods 21–25
Region 5: Periods 27–30

Alternatively, using Equation 9.33 yields $ARL = 9 \times 100 = 900$, using $k = 0.5$ and Equation 7.4:

$$h \approx \frac{904}{902} \ln(451) - 1.166 = 4.96 \tag{9.34}$$

This is very similar to the threshold value of 4.95 found when using Equation 9.32. These thresholds ensure that the average time until the first false alarm over the set of m charts is equal to $ARL = 100$.

This type of adjustment is appropriate and will yield the desired ARL when (1) no spatial autocorrelation in the regional variables exists, (2) all regions are in control, and (3) there is a desire to monitor individual regions, and not neighborhoods around regions. However, Equations 9.32 and 9.33 together with Equation 7.4 will often lead to thresholds that are too conservative (i.e., thresholds that are too high). One reason for this is that not all m regions may be "in-control"; we only require a threshold and false-alarm rate that have been adjusted for the number of in-control regions (which is unknown, but is less than or equal to m). When a region goes out of control, other (e.g., surrounding regions) may simultaneously go out of control. This suggests that the adjustments for multiple testing may be too severe, and recent developments in the area of multiple testing can be used to lower the thresholds (for a review, see Castro and Singer 2006). A second reason that Equations 9.32 and 9.33 can be conservative is that they assume that the m regional charts are independent. More commonly, regional charts may exhibit spatial dependence; a cusum chart for one region may look a lot like a chart for a nearby region. Finally, if emergent clusters might exceed the size of regional subunits, this will provide a rationale for monitoring local statistics for neighborhoods around regions.

Maintaining separate charts for each region is a *directional* scheme; the approach will work very well when the actual change occurs in one of the regions but can lose considerable power in detecting change quickly when changes in other directions occur. If, for example, an increase occurs in a neighborhood containing several regions (corresponding to several charts), this approach will not be as effective, and can yield longer times to detection than other methods.

9.2.2 Example: Kidney Failure in Cats

We now introduce a spatial element into the illustration involving kidney failure in cats (see Chapter 7, subsection 7.5.1.1). For simplicity, assume that the country had been divided into two regions of equal size. We now wish to monitor two geographic regions separately (perhaps pet food is manufactured and distributed separately within the two regions). We should now be able to detect change that occurs only within one of the regions more quickly. For comparison, we will need to maintain an identical false-alarm rate for the entire system. We have $\lambda_{0,i} = 1.15$ and $\lambda_{1,i} = 1.5$, each replacing the overall rate of $\lambda_0 = 2.3$ and the value of $\lambda_1 = 3.0$ used in Chapter 7, respectively, for each of the two regions ($i = 1, 2$), and this implies that $k = 1.317$ in each region. A threshold of $h = 15$ would yield an ARL_0 of about 1650 in each region, but this would lead to signaling more frequently than desired in at least one of the two regions. Consequently, we ask what threshold would yield an ARL of approximately $2 \times 1500 = 3000$ in each region; simulation reveals that $h = 17$ gives a value that is "close enough" at $ARL_0 = 2800$. If we maintain a cusum chart for each region with these parameters, simulation shows that we would detect an increase that occurred in one of the regions to $\lambda_1 = 1.85$ (while the other region's rate remains at 1.15, so that the overall rate for the system has risen to 3.0) in an average of 32 days, substantially shorter than the 41 days found in Chapter 7, subsection 7.5.1.1.

In practice, we are often unsure about what spatial scale may be the most appropriate—that is, at what scale we might expect observations to deviate from expectations. We may wish to implement both schemes (that is, the maintenance of two separate charts *and* a single, systemwide chart) simultaneously. If we did so, we would again have a multiplicity issue—although each scheme would by itself have the desired false-alarm rate of about once every 1500 days, the rate at which false alarms would occur in *at least* one of the two schemes would be higher. One solution to this is to use a common threshold for both schemes. Using $h = 17$, the threshold for the regional charts leads to a false alarm on at least one of the spatial scales every $ARL_0 = 950$ periods. Experimentation shows that $h = 19$ yields a desirable value of $ARL_0 = 1600$. A jump in the rate in one of the two regions to $\lambda_1 = 1.85$ is detected under this multiscale monitoring scheme in an average of 34 days, just slightly longer than the 32 days found earlier when only the regional charts are maintained.

A (usually small) drawback associated with using a common threshold across multiple scales is that it is biased against finding changes that occur at larger spatial scales. Generally, we will want to use lower thresholds when monitoring larger spatial scales; the effect of smoothing the data over a large region is that we do not need to raise the threshold as much on account of the multiplicity of regional testing because we have effectively fewer independent tests when monitoring on large spatial scales. Thus, we might consider raising the ARL_0 in each of the two schemes (in this case from about 1500 to some higher value that is common to both) to yield (separate) thresholds that imply the desired overall false-alarm rate. In the present example, the discreteness of the Poisson distribution yields the same threshold for the schemes at each of the two spatial scales (in both cases, $h = 17$ yields the desired ARL_0 of about 1500), and so separate thresholds are not necessary.

9.2.3 Example: Breast Cancer Mortality in the Northeastern United States

In this section we use *GeoSurveillance* to illustrate how several regions (in this case counties) can be monitored independently and simultaneously, using a Bonferroni adjustment.

After opening *GeoSurveillance*, choose Cluster Detection > Prospective test > Cusum for Normal Variate (Univariate) > Polygon (ESRI Shapefile), and then choose NEBreast.shp provided as a sample data set with the software.

This is the file associated with the breast cancer mortality data for 217 counties in the northeastern United States. At this point, a Select Name Fields dialog box will open. Click on OK to select the default choice for the primary ID. The map of the northeastern United States will open.

We will maintain 217 separate cusum charts, one for each county. This is equivalent to looking for changes in breast cancer mortality at the scale

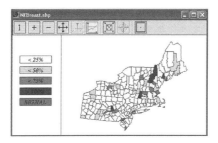

FIGURE 9.3
County cusum values in 1988 (year 21).

of the county. We will use a Bonferroni adjustment to account for the fact that we are maintaining so many charts simultaneously.

Under Tool, choose Cusum Tool. Leave $k = 0.5$ as it is, change the number of regions to 217, change sigma to 0, and enter 100 as the value for ARL_0. Choose Bonferroni correction under ARL_0 options; this leads to a threshold value of 8.124. (The reader could verify that this is the threshold associated with an ARL of $217 \times 100 = 21,700$ when maintaining a single chart.) After closing this Get h-value from ARL box, enter 8.124 into the h-value field under Parameters. Finally, in the Data option box, choose Fields selection and choose CS68 and CS88 in the drop-down boxes for from and to, respectively. These are the fields in the data set that contain the 217 z-scores for each year, from 1968 to 1988. Then click Run.

The first signal is sent in year 21, by Philadelphia and Delaware counties in Pennsylvania, and Ontario County in New York. Figure 9.3 portrays the county cusums in the 1988 (year 21); shading on the map reflects the cusum as a percentage of the threshold. Clicking on any of the individual counties displays the cusum chart for that county as shown in Figure 9.4 for Philadelphia County.

In the next section, we examine some alternative approaches to multiplicity adjustment.

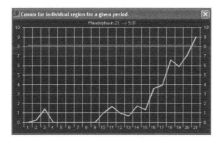

FIGURE 9.4
Cusum chart for Philadelphia County, Pennsylvania.

9.3   Monitoring Many Local Statistics Simultaneously

Now suppose that we wish to carry out surveillance of several local statistics simultaneously. As a reminder, we desire rapid detection of the shift from the null hypothesis (where there is no spatial pattern, and all regions have zero means) to the situation where a spatial cluster consisting of adjacent regions witnesses a change from regional means of zero to alternative, higher regional means.

We assume that each region i has a normally distributed observation x_{it} during each time period t. This is not particularly limiting because, in many instances, Poisson counts may be accurately transformed to normality for use in a cusum setting (Rossi et al. 1999). We thus assume that the x's have been standardized to have mean zero and variance one.

We could either keep a Raubertas-type chart for each local statistic, or, more generally (as it is possible to account for underlying spatial autocorrelation in the regional values), keep a Healy-type chart for each region. Consider first the special case where $\Sigma = I$; the Healy and Raubertas charts will be identical. An important issue is the adjustment for multiplicity; using individual thresholds for each chart based upon mARL, where m is the number of regions, would be too conservative because the charts will be correlated (nearby local statistics will be similar as there is some commonality in the regional values they make use of). On the other hand, thresholds based on ARL alone would be too liberal, unless the charts for all local statistics were identical. It is of interest to find the number of effectively independent charts (say, e); in that case, each individual threshold could then be based upon e(ARL).

Let the regional variables be denoted by $\{y_i\}$ and the local statistic be monitored by $\{z_i\}$. One alternative for the determination of appropriate thresholds would be Monte Carlo simulation; a value of $e < m$ could be used in $\{(e)\text{ARL}_0\}$, and the value of e that leads to a systemwide average run length of ARL_0 could be found via simulation.

Alternatively, we can use a Gaussian kernel to define the neighborhood weights, as suggested in Chapter 5, Section 5.10:

$$z_i = \sum_j w_{ij}' y_j$$

$$w_{ij}' = \frac{w_{ij}}{\sqrt{\sum_j w_{ij}^2}}$$

$$w_{ij} = \frac{1}{\sqrt{\pi}\sigma} \exp\left(\frac{-d_{ij}^2}{2\sigma^2(A/m)}\right) \tag{9.35}$$

where A is the size of the study area, and σ is the width of the Gaussian kernel, expressed in terms of multiples of the square root of the average regional area. As noted in Chapter 5, these local statistics, z, have normal distributions with mean zero and variance one (e.g., see also Siegmund and Worsley 1995; Rogerson 2001b).

Then, one possibility that accounts for the correlation among the local statistics is to use the effective number of independent regions or resels discussed in Chapter 5, subsection 5.10.5, as an estimate of e. The equation for this is repeated here for convenience:

$$e = \frac{m}{1+0.81\sigma^2} \tag{9.36}$$

To assess the feasibility of this idea, a set of simulation experiments was designed to evaluate whether this is of use in a monitoring context. A 16×16 grid of cells was filled with normal standard deviates for each successive time period, and then smoothed with a Gaussian kernel using various values of σ. The central 8×8 portion of the grid was then taken as the study area to avoid possible edge effects. Cumulative sums were kept for each of the 64 local statistics (i.e., for each of the y_{it}; $i = 1, 2, …, 64$).

Results are shown in Table 9.4. The first two columns give the parameters chosen for particular simulation runs. For each pair of σ and h values shown in the table, 200 trial runs were carried out, and the time until the first out-of-control signal (indicating that the mean of at least one local statistic was now greater than zero) was recorded. To speed the simulations,

TABLE 9.4

Effective Number of Independent Tests When Monitoring Smoothed, Neighborhood z-Scores

(1) σ	(2) h	(3) Estimated ARL_0 from Simulations	(4) ARL_0 Based on h and One Region	(5) Effective Number of Indep. Tests	(6) Estimated Number of Indep. Tests (Equation 9.36)	(7) h Determined Using Columns (3) and (6)
1	5.5	55.8	1,555	$1555/55.8 = 27.9$	35.4	5.74
1	6.0	81.7	2,573	$2573/81.7 = 31.5$	35.4	6.12
1	6.7	137.0	5,196	$5196/137 = 37.9$	35.4	6.63
1	7.4	277	10,481	$10481/277 = 37.8$	35.4	7.33
1	8.0	433.5	19,112	$19112/433.5 = 44.1$	35.4	7.78
1	8.7	960.5	38,506	$38,506/960.5 = 39.3$	35.4	8.58
1	9.2	1334	63,499	$63,499/1334 = 47.6$	35.4	8.90
2	4.5	52	564	$564/52 = 10.8$	15.1	4.83
2	5.5	111.5	1555	$1555/111.5 = 13.9$	15.1	5.58

censoring points were chosen; if the number of time periods needed for a false alarm exceeded a value of C (where the value of C was particular to each pair of σ and h values), this was noted and the ARL_0 was estimated using the fact that average run lengths tend to have an exponential distribution (see the appendix to this chapter).

The third column gives the estimated ARL under the null hypothesis and is based upon the simulations. The fourth column gives the ARL that would result if the corresponding value of h was used in a cusum with only one region. The number of effectively independent tests, as determined from the simulation, is shown in the fifth column; it is found by dividing column (4) by column (3). This is to be compared with column (6), which is the estimated number of independent tests based upon Equation 9.36. In general, columns (5) and (6) are quite similar. For low values of ARL_0, there is a tendency for the estimated number of independent tests to be too high (i.e., conservative). For large values of ARL_0, the opposite is true— the estimated number of independent tests is too low, and this results in a somewhat higher number of false alarms than desired.

This comparison can also be viewed by comparing columns (1) and (7); the latter is the value of the threshold h that would be used if one used Equation 9.36 to estimate the number of effectively independent tests, and if one desired an ARL_0 equal to that given in column (3). Again, low values of ARL_0 would results in the use of thresholds (h) that were conservative; that is, the thresholds would be too high (in comparison with the h values of column (2), which are the ones that *should* be used to achieve the ARLs in column (3)). In contrast, for high values of ARL_0, the estimated values of h are lower than the values in column (2) that in principle should be used to achieve the ARLs in column (3).

Although this idea gives results that are similar to those found through Monte Carlo simulation, the adjustment is based on the (static) correlation between regional local statistics observed at a single point in time. In practice, the cusum charts being maintained for each regional local statistic will have correlations that are not necessarily the same as this static correlation. Any adjustments to chart thresholds should in principle be based upon the probabilities of charts jointly signaling. Additional approaches to monitoring data from multiregional systems include methods designed for multivariate surveillance (Rogerson and Yamada 2004a) and monitoring regional maxima (Rogerson 2005b).

One more illustration of monitoring local statistics is now made by reconsidering the data in Table 9.2. From Equation 9.35 with $\sigma = 1$, the local statistic constructed for the center square would have weights of .2119 for corner squares, .3421 for squares adjacent by rook's adjacency, and a weight of 0.576 for the central region (region 5). The results of using a cusum for this weighted local statistic are shown in Table 9.5.

The cusum exceeds the threshold of 2.84 used for a single variable in time period 18 and time periods 20–30. It exceeds the conservative threshold of

TABLE 9.5

Cumulative Sum for the Local
Neighborhood Statistic of the
Center Square (subregion 5) in the
Hypothetical Nine-Region System
Presented in Figure 9.2

Time	Cumulative Sum for Region 5
1	0
2	0
3	0
4	0
5	0
6	0
7	0
8	0
9	0
10	0.65
11	0
12	0
13	0
14	0
15	1.85
16	1.36
17	2.59
18	3.07
19	2.26
20	3.05
21	3.67
22	3.9
23	5.39
24	5.77
25	6.59
26	6.69
27	6.26
28	6.53
29	6.56
30	5.89

4.96 associated with monitoring nine independent local statistics in time
periods 23–30.

9.3.1 Example: Breast Cancer Mortality in the Northeastern United States

We now illustrate the search for geographic clusters of change that might
occur on a scale greater than that of the county (in contrast with subsection 9.2.3, where we searched for increases at the scale of the county; it

may be, for instance, that increases in breast cancer mortality occur across a set of contiguous counties). Now under Tools, choose Cusum Tool once again, and use $k = 0.5$, regions = 217, $\sigma = 1$, and $ARL_0 = 100$. The value of $\sigma = 1$ is associated with the concept of centering a normal distribution on a county, where the normal distribution has a standard deviation equal to the average distance between county centroids. Under ARL_0 options, choose Spatial autocorrelation, and Get Result. The appropriate threshold is shown to be 7.534. Enter 7.534 into Parameters, and change sigma from the default of 0 to a value of 1, and click Run. Now the first signal is sent in year 17, by Franklin and Hampshire counties (adjacent counties in Massachusetts). These, and only these counties, continue to signal in years 18–20. In the last year, a total of five counties exceed this threshold (including Franklin, but not Hampshire). Figure 9.5 shows the cusum map in 1988; note the geographic clusters in Massachusetts, New Hampshire, and the Philadelphia area; these regions have accumulated a relatively high number of cases of breast cancer mortality relative to expectations.

9.3.2 Poisson Variables

The foregoing has assumed that regional variables come from a normal distribution. For Poisson variables, one can monitor the quantities $y_{i,t} = \sum_j w_{ij} x_{j,t}$, where $x_{j,t}$ is the observed count in region j at time t, and w_{ij} is a weight associated with, for example, the distance from region i to region j. These observed quantities are then compared with their corresponding expectations $\sum_j w_{ij} \lambda_{0,t,j}$ (where the subscript t, j refers to region j at time t), and used in a cusum for region i.

To determine appropriate critical thresholds for each region, Monte Carlo simulation of the null hypothesis may be used (where observed counts are realizations from Poisson or normal distributions with parameters set equal to the corresponding expectations). In particular, with a desired average run length of ARL_0, the critical thresholds should be determined using $e(ARL_0)$; the value of e is less than the number of regions (m), and

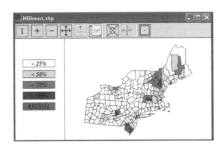

FIGURE 9.5
County cusum values in 1988 using Gaussian local statistic with $\sigma = 1$.

is determined via simulation to lead to the desired average run length. The greater the correlation between the local regional statistics, the lower e will be relative to m.

9.4 Summary

The prospective surveillance of geographic patterns, based on incoming streams of spatial data, is a field that has grown rapidly in the last decade. This growth has been motivated largely through interest in public health surveillance. It should be realized, however, that there are also many potential applications in other areas, including applications to crime analysis, where there is interest in emerging areas of criminal activity, and in marketing, where the spatial pattern of customers' shopping behaviors in a competitive retailing environment could be monitored.

Appendix

To find r in terms of u, we first write

$$g(r) = u - \frac{\theta_1 + \theta_2 r^2 - \theta_1 e^{-\theta_2 r^2}}{\theta_1 + \theta_2 R^2 - \theta_1 e^{-\theta_2 R^2}} = 0$$

Taking the derivative, we find

$$g'(r) = \frac{-2\theta_2 r(1 + \theta_1 e^{-\theta_2 r^2})}{\theta_1 + \theta_2 R^2 - \theta_1 e^{-\theta_2 R^2}}$$

An initial guess for r, say r_0, is made. This initial guess is updated to derive an improved estimate of r, say r_1, via

$$r_1 = r_0 - \frac{g(r_0)}{g'(r_0)}$$

This is used to iterate until convergence has been achieved.

10

More Approaches to the Statistical Surveillance of Geographic Clustering

10.1 Introduction

In this chapter we examine two alternative approaches to spatial surveillance.

The next section (10.2) is devoted to the question of determining when the peak (or maximum) on a series of maps becomes sufficiently high that an alarm is sounded. For example, peaks on maps of disease rates will, of course, fluctuate over time, and it is of interest to be able to determine if and when the peaks are getting higher, in comparison with some baseline norm for such peaks. The Gumbel distribution is used as a model for the statistical distribution of extreme values; this distribution does not require the underlying distributions of regional values to be either normal, known, or identical. Cumulative sum surveillance methods are used to monitor these Gumbel variates, and these methods can also be extended for use when monitoring smoothed regional values (where the quantity monitored is a weighted sum of values in the immediate geographical neighborhood). The methods are illustrated by using several cancer mortality datasets for the United States.

The following section (10.3) explores the applicability and feasibility of using multivariate surveillance in a geographic context. Multivariate surveillance, as the name implies, focuses upon the goal of simultaneously monitoring many variables. This was also the aim of the methods discussed in Sections 9.2 and 9.3. Each of the variables corresponds to an observation for a region; there are as many variables as there are regions. This section is based largely on Rogerson and Yamada (2004a), who compare the multiple-region univariate approach (using a Bonferroni adjustment) from Chapter 9 with multivariate approaches to spatial monitoring. The multiple-region univariate method is not an explicitly spatial approach because individual regions are monitored independently with no spatial relationship between them taken into consideration. As detailed later in this chapter, the multivariate approach accounts for the covariation between regional observations but is still a "global" method in the sense that a single multivariate distance measurement is used to assess the deviation between observation and expectation. Thus, particular local spatial clusters are not being monitored. To carry out multivariate surveillance in a general way, however, it is necessary to specify the variance–covariance matrix, and this makes implementation of multivariate surveillance in a geographical context more challenging.

10.2 Monitoring Spatial Maxima

There is often interest in assessing the extreme values on maps displaying regional data (e.g., see Stern and Cressie 1999). For instance, a state health official may be interested in the peak rate on a map displaying county-level mortality or morbidity rates. Similarly, a crime analyst for a metropolitan police department may wonder how to react to the maximum value on a map of crime rates by ZIP code. All such maps have peaks, but whether or not the peaks are statistically significant to warrant investigational efforts is a different problem. Are the peaks high enough to reallocate police enforcement effort? Does the health official declare a public health investigation in the region where mortality and morbidity rates are at their highest? It is important to be able to distinguish between "statistically significant" maxima and those peaks that are merely the result of random fluctuation and thus uninteresting. Scan statistics as described in Chapter 5 are designed to answer questions of this type.

If the crime analyst observes such maps every day, how should a desire to minimize the number of unnecessary investigations be balanced against the need to detect new, significant peaks as quickly as possible? In this section we will review methods for deciding when the maxima that appear on a temporal series of regional maps become sufficiently great that we can no longer assume that they are simply the typical fluctuations of a "null" process (for example, the "null" process of equal disease risk will, by chance alone, produce map peaks that vary in height over time).

We will assume that data have been collected for a set of regional units. In some cases we will use the data for all regional units directly; in other cases, we will be interested in the smoothed maps that are constructed from the local statistics for each region. In the latter case, the local statistic in each region will consist of a weighted sum of the regional values in a neighborhood of the region of interest. We will be interested in assessing the significance of the map's peak, and whether the peaks on a recent series of maps are significantly higher than what was expected.

The primary purpose of this section is to outline approaches to monitoring the maximum of a set of regional values that are observed over time. The section is structured as follows. In subsection 10.2.1, approaches for monitoring a temporal sequence consisting of the maximum of a set of regional values are discussed. In particular, the Gumbel distribution is used to model the maximum value; this has the advantage of not necessarily requiring stringent assumptions on the distribution of underlying regional values. In subsection 10.2.2, these methods are extended and discussed in the context of monitoring a smoothed surface of regional values, where the local statistics of each region are a weighted sum of the values in the immediate geographical neighborhood. In subsection 10.2.3, data on breast cancer mortality for the 217 counties of the northeastern United States, and prostate cancer mortality for the entire United States, both during the period 1968–1998, are used as an illustration. The final subsection (10.2.4) provides a summary.

10.2.1 Monitoring Spatial Maxima

Assume that we observe regional values on a map for successive time periods $(x_{it}, i = 1,2, \dots ,m; t = 1,2, \dots)$. We are interested in monitoring the maximum regional value observed during each period, and there is particular interest in the null hypothesis that the distribution of these extreme values has not changed over time. The alternative hypothesis is that there is a point in time when the distribution of extreme values has changed and is characterized by higher extreme values, implying, for example, increased risk of disease incidence or crimes. Under this alternative, and after standardization for different population sizes in the different regions, the maximum observed value should be sensitive to an outbreak anywhere within the regions.

It will be convenient to work with standardized scores for each region; these may be constructed in several ways, using the observed and expected values (including, e.g., the Rossi et al. and the Freeman–Tukey transformations described in subsection 3.3.7, and a Poisson transformation, achieved by subtracting the expected value from the observed, and then dividing the result by the square root of the expected value).

Use of standardized scores allows direct comparison of the observed values from regions with different expectations—for example, due to different at-risk populations. Consider two regions—one with an expectation of 9 cases, and the other with an expectation of 100 cases. Suppose that 15 cases are observed in the first region ($z = 1.87$ using the Rossi et al. transformation), and 167 cases are observed in the other region ($z = 6.27$, again using the Rossi et al. transformation). The ratio of observed to expected cases is 1.67 for both regions. We expect more variability in this ratio in the region with the smaller expectation and, as expected, its z-score is less extreme. The z-scores are derived to yield associated tail probabilities that reflect the likelihood of a number of cases equal to or greater than what is observed, under the null hypothesis of uniform risk. In this case, the tail probability associated with $z = 6.27$ is very small, reflecting how unusual 167 cases would actually be if we only expected 100.

10.2.1.1   Type I Extreme Value (Gumbel) Distribution

The Type I extreme value, or Gumbel distribution, characterizes the relative frequency distribution of the maximum of a large number of observed variables that each come from one of many possible underlying distributions, including the normal (Gumbel 1958). Its cumulative distribution function is

$$F(y) = e^{-e^{-\alpha(y-u)}}; \quad -\infty \le y \le \infty \tag{10.1}$$

The mean of the Gumbel distribution is $u + .577/\alpha$; the standard deviation is $1.282/\alpha$. The parameter u is the mode of the distribution.

An observed sequence of observations can be used to estimate the parameters using the method of moments as follows. For a sequence of observations $(y_1, \dots, y_t)$, the observed and theoretical means and standard deviations may be equated to yield

$$u = \bar{y} - \frac{.577\sigma_y}{1.282}$$

$$\alpha = \frac{1.282}{\sigma_y} \tag{10.2}$$

Bollobas (1985) shows that, in the special case with m normally distributed variables, the cumulative distribution function of the maximum, Y, is approximately as follows:

$$\Pr(Y < y = a_m x + b_m) = e^{-e^{-x}} \tag{10.3}$$

where

$$a_m = \frac{\sqrt{2\sigma^2 \log m}}{2 \log m}$$

$$b_m = \mu + \sqrt{2\sigma^2 \log m} \left(1 - \frac{\log \log m}{4 \log m} - \frac{\log(2\pi^{1/2})}{2 \log m} \right) \tag{10.4}$$

For example, let $m = 30$, $\mu = 0$, and $\sigma^2 = 1$. Then, $a_m = .3834$, and $b_m = 1.888$. Suppose we observe a maximum of $y = 2.27$. This implies $x = 1$ (because $x = (y - b_m)/a_m$), and $\Pr(Y < 2.27) = .6922$. Similarly, the probability of observing $Y < 2.655$ is .873 because $y = 2.655$ is associated with $x = 2$. The values of a and b may be related to the parameters of the Gumbel distribution using $\alpha = 1/a_m$ and $u = b_m$. For this example, $\alpha = 1/.3834 = 2.608$, and $\mu = 1.88$. A simulation using $m = 30$ standard normal random variates and 10,000 Monte Carlo repetitions yielded $\bar{y} = 2.045$ and $\sigma_y = 0.495$. This leads to $u = 1.822$ and $\alpha = 1/.3861$, and these are close to the values just found.

10.2.1.2 *Cusum Surveillance of Gumbel Variates*

Hawkins and Olwell (1998) describe a general strategy for formulating optimal cusum schemes for variables from the exponential family of distributions. Because the Gumbel is a member of this family, here we outline a strategy for maintaining a cusum of Gumbel distribution variables.

Members of the exponential family of distributions may be written as

$$f(y|\theta) = \exp\{a(y)b(\theta) + c(y) + d(\theta)\} \tag{10.5}$$

Hawkins and Olwell note that, for this family of distributions, a cusum scheme monitors the quantity $a(y)$, with

$$k = -\frac{d(\theta_1) - d(\theta_0)}{b(\theta_1) - b(\theta_0)} \tag{10.6}$$

where θ_0 and θ_1 are the parameter values for the in- and out-of-control scenarios, respectively. For the Gumbel distribution, we may write

$$f(y|u) = \alpha \exp\{-\alpha y + \alpha u - \exp(-\alpha y + \alpha u)\} \tag{10.7}$$

The parameter u is the mode of the distribution, and the parameter α is related to the dispersion of the distribution.

We assume that the parameter α is fixed and that we wish to carry out surveillance on the parameter u. Thus, in terms of Equation 10.5, we have

$$a(y) = -\exp(-\alpha y)$$

$$b(u) = \exp(-\alpha u)$$

$$c(y) = -\alpha y$$

$$d(u) = \alpha u \tag{10.8}$$

Therefore,

$$k = \frac{\alpha(u_1 - u_0)}{e^{\alpha u_1} - e^{\alpha u_0}} \tag{10.9}$$

Using the previous example with $m = 30$, $a_m = .3834 = 1/\alpha$, and $b_m = u_0 = 1.888$, let $u_1 = 2.3$ (that is, we wish to minimize the time it takes to detect an increase in the mode of the distribution of the maxima from 1.888 to 2.3). Then, $k = 0.00405$. One approach to determining the threshold h is to simulate the null hypothesis. This is achieved by choosing 30 standard normal deviates and selecting the maximum. These maxima (y) are then used in the cusum:

$$S_t = \min(0, S_{t-1} + e^{-\alpha y_t} - k)$$

or

$$S_t' = \max(0, S_{t-1} - e^{-\alpha y_t} + k) \tag{10.10}$$

If, for example, we desire ARL = 100, we try different thresholds (h) until we find one that is consistent with our desired ARL. Here, we find $h = -.0106$ (or, equivalently, $h' = .0106$). An alarm is sounded when $S_t < h$ (or equivalently, when $S_t' > h'$).

10.2.1.3 Example: Female Breast Cancer Mortality Rates in the Northeastern United States

Previous studies of female breast cancer mortality in the northeastern United States have either focused upon a cross-sectional assessment of geographic clustering for one period of time (Kulldorff et al. 1997), or have examined temporal change from a retrospective view (Han and Rogerson 2003). In the latter case, annual data for the period 1968–1998 are described and statistically significant changepoints are found when the geographic pattern of breast cancer shifts.

From the Compressed Mortality File (CMF; National Center for Health Statistics), annual, county-level data on breast cancer mortality were extracted for the 217 counties comprising the northeastern United States for the period 1968–1998. The expectations for each county and for each year were age adjusted; they were computed from nationwide age-specific breast cancer rates and the age-specific county populations for each year provided in the file. Thus, changes in population size and age structure were taken into account when deriving expectations. Standardized z-scores were derived using the Rossi transformation, which yields z-values that are consistent with how likely or unlikely the observations are, given the expectations. The maximum z-scores for each year, together with the county they occurred in, are shown in Table 10.1.

We now illustrate how these maxima could have been used in a surveillance framework. The most straightforward approach would be to first choose a desired ARL and then use a Shewhart chart. If ARL = 100, then $\Phi^{-1}(.01/217) = z^* = 3.91$. This is exceeded twice—in 1972 and 1973, by counties in Pennsylvania and Vermont, respectively.

There are two ways to implement a cusum scheme. One is to assume that the null distribution of the maximum is based on the maximum of 217 z-scores, as the Shewhart chart does. Using Equation (10.4) with $m = 217$, yields $\alpha = 3.28$ and $u = 2.638$. This was corroborated by repeatedly simulating 217 standard normal deviates and choosing the maximum among them, and led to similar parameter estimates of $\alpha = 3.28$ and $u = 2.596$ using the method of moments. Using the latter parameter values and minimizing the time to find an increase in the mode of this distribution from $u = 2.596$ to $u_1 = 2.9$ yields $k = .000148$; simulation of the null hypothesis yields a threshold of $h = .00046$ for ARL = 100. When surveillance commences in 1968, the cusums for the first 7 years are shown in Table 10.2. The cusum approaches, but does not quite reach, the critical

TABLE 10.1

Maximum z-Scores and Their Locations:
Female Breast Cancer Mortality in the
Northeastern United States, 1968–1998

Year	County	Maximum z-score
1968	New York, NY	3.17
1969	New York, NY	3.75
1970	Nassau, NY	3.06
1971	Nassau, NY	3.22
1972	Clinton, PA	4.14
1973	Grand Isle, VT	3.99
1974	Yates, NY	2.28
1975	Nassau, NY	2.98
1976	Nassau, NY	2.89
1977	Orleans, VT	2.62
1978	Onondaga, NY	3.53
1979	Erie, NY	2.69
1980	Nassau, NY	2.79
1981	New York, NY	2.78
1982	Monmouth, NJ	3.78
1983	Nantucket, MA	3.16
1984	Plymouth, MA	2.30
1985	Norfolk, MA	2.69
1986	Hunterdon, NJ	2.63
1987	Orleans, VT	2.77
1988	Albany, NY	2.52
1989	Albany, NY	2.19
1990	Philadelphia, PA	2.42
1991	Erie, NY	2.44
1992	Camden, NJ	2.42
1993	Philadelphia, PA	2.32
1994	Essex, NJ	2.94
1995	Beaver, PA	2.49
1996	Herkimer, NY	2.57
1997	Philadelphia, PA	3.09
1998	Philadelphia, PA	1.97

value in 1973. From that point on (data not shown), the cusum decreases and remains at or near zero; again, no departure from the null hypothesis of no change is indicated.

Alternatively, the cusum approach could be implemented by using the first several years of the dataset to establish a baseline. This approach is more model-based; it uses a historical baseline to establish the parameters of the extreme-value distribution but does not strictly assume that the distribution necessarily represents the maximum of m normally distributed variables. Here, we used the first 10 years for the baseline; the mean maximum value is 3.31, and the standard deviation of these 10

TABLE 10.2

Cumulative Sum (Cusum) Results
for Breast Cancer Mortality Data
Shown in Table 10.1

Year	Cusum
1968	.000135
1969	.000098
1970	.000111
1971	.000120
1972	.000262
1973	.000369
1974	0

Note: Critical value is $h = 0.00046$.
Cusum remains at zero after 1974.

values is 0.4941. Using Equation 10.2, we estimate the parameters of the Gumbel distribution and find that $u = 3.088$ and $\alpha = 2.595$. Suppose we are interested in finding an increase in the mode of the distribution of the maxima from $u = 3.088$ to $u_1 = 3.3$ as quickly as possible. Then Equation 10.9 yields $k = .0002488$. If we choose a false-alarm rate of one every 100 years (ARL = 100), Monte Carlo simulation of the null hypothesis (carried out by choosing random deviates from a Gumbel distribution with the parameters u and α) yields $h = 0.0011$. Using these in the cusum equations (Equation 10.10), with surveillance beginning in the 11th year (1978), the cusum remains below the critical value of h during the entire surveillance period (i.e., through 1998). This implies that the maxima are not becoming more extreme over time and accords with a cursory glance at Table 10.1, which reveals that the magnitudes of the maxima do not appear to be increasing.

10.2.1.4 *Example: Prostate Cancer Data in the United States*

Annual, county-level data on male mortality from prostate cancer were also examined, this time for the entire United States. The maximum z-score observed in each year is shown in Table 10.3 (among those counties with expectations of more than one case in that year). Using the first 10 years as the base period, the mean value of the maximum for this period is 3.90, and the standard deviation is 0.647. These values lead to Gumbel parameter estimates of $\alpha = 1.982$ and $u = 3.606$ (using Equation 10.2). Minimizing the time it takes to find a change from $u_0 = 3.606$ to $u_1 = 4.0$ leads to $k = .000519$ (using Equation 10.9). A Monte Carlo simulation using $ARL_0 = 100$ resulted in $h = 0.0019$. The cusum is started in 1979, and crosses the threshold in 1989. It remains over the threshold thereafter, indicating that the distribution of maximum values has likely changed relative to the base period.

TABLE 10.3

Maximum z-Scores and Cusum Values for Prostate Cancer
Mortality Data for the Entire United States, 1968–1998

Year	Maximum z-Score	Cumulative Sum (10-year base period)	Cumulative Sum (9-year base period) ($\times 10^{-8}$)
1968	4.16		
1969	3.52		
1970	3.70		
1971	3.67		
1972	—		
1973	3.93		
1974	4.16		
1975	4.99		
1976	3.60		
1977	4.12		
1978	5.69		3.6
1979	4.11	0.00023	6.2
1980	3.78	0.00019	5.7
1981	4.75	0.00026	6.3
1982	4.40	0.00011	3.3
1983	3.51	0	0
1984	4.47	0.00038	3.4
1985	5.98	0.00072	6.7*
1986	4.74	0.00116	10.2
1987	4.14	0.00140	12.9
1988	4.36	0.00175	16.2
1989	5.97	0.00207*	19.4
1990	4.60	0.00248	22.9
1991	4.20	0.00276	25.8
1992	4.16	0.00301	28.6
1993	4.03	0.00319	30.8
1994	4.15	0.00317	33.5
1995	4.27	0.00348	37.0
1996	4.63	0.00388	40.5
1997	4.57	0.00429	43.9
1998	4.67	0.00468	

* Indicates first year threshold is exceeded.

Note: Critical thresholds: 10-year base period, 0019; 9-year base period, 6.6×10^{-8}.

One difficult issue pertains to the choice of the base period. In the case of prostate cancer mortality, if the base period had consisted of the first 9 years instead of the first 10, the cusum would have signaled 4 years sooner, in 1985 (and would have almost signaled in the second and fourth year of monitoring); see the fourth column of Table 10.3. One alternative is to construct a self-starting cusum; these are discussed in Chapter 7, subsection 7.7.2, and in more detail by Hawkins and Olwell (1998).

10.2.2 Determination of Threshold Parameter

In the previous section, Monte Carlo simulation was used to derive the critical threshold h. In this section, we show that it is possible to avoid the use of simulation by deriving analytically a value for the threshold. The approach is based upon the fact that when y_t comes from a Gumbel distribution with parameters α and u, the quantity $e^{-\alpha u}$ has an exponential distribution with parameter αu and mean $e^{-\alpha u}$.

We draw upon the discussion of the exponential cusum in Chapter 7, Section 7.6, where interest lies in detecting an increase from an in-control exponential parameter θ_0 to an out-of-control value $\theta_1 > \theta_0$. This increase in the parameter corresponds to a decrease in the mean of the exponential variable.

To illustrate how this may be applied in the context of surveillance of maxima, consider the application to the prostate cancer data described in the previous section. With a 10-year base period, the parameters were $\theta = 1.982$ and $u_0 = 3.606$, and the objective was to minimize the time to detect a change to $u_1 = 4.0$. The quantity $x_t = e^{-1.982y_t}$ has exponential parameter $\theta_0 = e^{1.982(3.606)} = 1270$ when the process is in control. The out-of-control exponential parameter θ_1 is equal to $e^{1.982(4)} = 2774$. The observed quantities $e^{-1.982y_t}$ are standardized by multiplying them by 1270; this gives the observed quantities an in-control exponential distribution with a mean of one. The out-of-control parameter is now $\tilde{\theta}_1 = \theta_1/\theta_0 = 2774/1270 = 2.18$. Equation 7.12 yields $\tilde{k} = \{2.18 - 1\}/\{\ln(2.18)\} = 1.514$. This, in turn, implies that $q = 100 \ln (2.18) |1 - 1.514| = 40.057$ and $h = 3.553$. Using the cusum in Equation 7.13 with these parameters yields results similar to those shown in Table 10.3 for the case of a 10-year base period.

10.2.3 Summary

The surveillance methods described in this section may be used to monitor the maxima observed among a series of regional values observed over time. Whereas other types of cusum surveillance may be appropriate when there is interest in monitoring individual regions (e.g., using the assumption that counts, rates, or transformed values are Poisson or normal), the use of the Gumbel distribution is appropriate for variables that represent maxima. The Gumbel distribution arises as the limiting form for the maximum among a number of random variables that have loosely specified tail distributions; it therefore does not require assumptions that the underlying regional values are normal, known, or identically distributed.

One question not addressed here is how to interpret a signal. When the cusum exceeds the threshold, it implies that the distribution of the maximum has likely changed. It will usually be of interest to identify a specific region or subset of regions that possess the maximum value in each of the time periods leading up to the signal.

Different alternative processes could lead to signals, and it will likely be difficult to discern what has given rise to any particular signal. For example, changes in screening practices in some regions but not others could lead to greater variability in cancer rates, causing the distribution of the maximum to change. A change in the distribution of the maximum could also be caused by the onset of a contagious disease, where the maximum value might likely occur in different regions at different time periods. It is precisely this type of situation where the use of the Gumbel monitoring approach might be most effective in detecting change. The methods could also be useful in syndromic surveillance, where there could be interest in the detection of small increases distributed across several regions during several time periods due to terrorism or other causes.

Finally the surveillance of extreme values could also be generalized in other ways. For instance, it would be of interest to monitor the minimum *p*-value observed on a set of maps over time. Under the null hypothesis of no raised incidence, maps of *p*-values should have a uniform distribution, with no spatial autocorrelation. It would be desirable to identify critical time points where the distribution of the minimum *p*-value changed because this would indicate the onset of potentially meaningful change.

10.3 Multivariate Cusum Approaches

10.3.1 Introduction

In this section, our objective is to compare several alternative approaches to spatial surveillance when data are collected periodically for more than one region. We focus on the comparison of a multivariate surveillance method with the methods discussed in that are based upon multiple univariate surveillance. We initially pay particular attention to describing the multivariate method because it has not previously been used in the context of public health surveillance in a spatial context. As Sonneson and Bock (2003) note, multivariate approaches to spatial surveillance constitute a desirable way to include information about the spatial structure of regional observations. Although multivariate surveillance methods have not been previously used in a spatial context, these authors note that the methods have been used in nonspatial applications to public health (Chen 1978; Chen et al. 1982).

We assume that data are in the form of observed regional counts, and there exists a set of expected regional counts associated with them. Expected regional counts might be based upon the age, sex, and race structure of the region, using indirect age standardization. More sophisticated models for expected counts could also be developed to account for other covariates (e.g., socioeconomic characteristics, month of the year, etc.) An example is provided in the work of Kleinman et al. (2004); they develop a generalized linear mixed model to estimate the probability that an individual visits a doctor's office on a given day.

In each of the approaches we examine, the objective is to detect as quickly as possible any deviations in the observed regional counts from the expected counts. Subsection 10.3.2 provides an introduction to alternative methods for the quick detection of changes in spatial patterns. In Section 10.3.3, we compare a multivariate method for surveillance (where variables each associated with an individual region are integrated into a single surveillance scheme) with the strategy of monitoring each region separately, using univariate surveillance in each region. In subsection 10.3.4 the alternatives are illustrated using annual data on breast cancer mortality for the northeastern United States, for the period 1968–1998. Subsection 10.3.5 provides a discussion of the methods and findings.

10.3.2 Alternative Approaches to Monitoring Regional Change for More Than One Region

In addition to the objective of monitoring more than one region at a time, it may be desirable to account for any covariation in the rates of nearby regions. In the following subsections, we review several alternative approaches to multivariate surveillance and assess their merits and limitations. These include the following:

(a) Separate monitoring of each regional disease count (Woodall and Ncube 1985)

(b) Monitoring outliers among vectors of the distance between observed and expected standardized counts, where the elements of vectors correspond to particular regions (e.g., a multivariate Shewhart chart)

(c) Monitoring a univariate measure of the distance between regional vectors of observed and expected standardized counts

(d) Monitoring the multivariate vector of cumulated differences between observed and expected standardized counts (Pignatiello and Runger 1990, Crosier 1988).

In our application of multivariate monitoring, each variable is taken to be a regional count. Pignatiello and Runger (1990) compare these four approaches when each has the same ARL_0 under the null hypothesis of no change. They find (d) to be the best in the sense that its ARL under the alternative hypothesis of a change in the mean (denoted ARL_1) is smallest when deviations from the expected regional counts begin to occur. When assessing the differences in two approaches by using the average run length under the alternative hypothesis as a comparative measure, it is helpful to keep in mind that the distribution of run lengths is approximately exponential.

When the number of variables (in our case, regions) is large, and the shift away from expectations is due to one region (say from $\mu_0 = \{0\,0\,0\,0\,...\,0\}$ to $\mu_A = \{\delta\,0\,0\,0\,...\,0\}$) they find method (a) to be slightly better than (d). However, method (a) can be sensitive to the *direction* of the shift. Suppose a shift of the same magnitude is instead composed of small changes in each region (say, from $\mu_0 = \{0\,0\,0\,0\,...\,0\}$ to $\mu_A = \{\delta/\sqrt{m}, \delta/\sqrt{m}, ...\,\delta/\sqrt{m}\}$, where m is again the number of regions or variables). Then the ARL_1 needed to detect the change with method (a) will be relatively longer.

As is well known, the multivariate Shewhart chart (method (b)) does not compare favorably with the other alternatives when the size of the deviation one wishes to detect is small. Crosier (1988) compares methods (c) and (d), and also finds that (d) is superior. The method for spatial surveillance described in Rogerson (1997) and in Chapter 8, section 8.4.1 is similar to method (c) because a univariate measure of spatial pattern (Tango's [1995] statistic) is monitored.

When regional covariation of rates is permitted, Pignatiello and Runger show that method (a) can give ARL_1's that vary substantially with the direction of the shift. Woodall and Ncube (1985) suggest monitoring the principal components instead of the original variables, but as Pignatello and Runger point out, this will not work well when a change occurs in an individual, original variable. Pignatiello and Runger's proposal to augment the number of variables monitored from m to some greater number does not generally work well because, although the augmentation results in better coverage of potential change directions, it loses power because of the augmentation.

In the next sections, we apply these methods (focusing upon (a), (b), and (d)) to both simulated data and data from the CMF (National Center for Health Statistics) to understand changes in the spatial pattern of breast cancer. Han and Rogerson (2003) find changes in the spatial pattern of breast cancer in the northeastern United States during the period 1968–1998. In particular, the cluster described by Kulldorff et al. (1997) in and near New York City appears to "migrate" and extend itself in the direction of Philadelphia during the period.

10.3.3   Methods and Illustrations

10.3.3.1   *Multivariate Monitoring*

The discussion in this section follows closely that of Pignatiello and Runger (1990). Multivariate monitoring is based upon the cumulative difference between the observed and expected numbers of cases:

$$\mathbf{S}_t = \sum_{j=t-n_t+1} (\mathbf{O}_j - \mathbf{E}_j) \tag{10.11}$$

where $\mathbf{O}_j$ and $\mathbf{E}_j$ are vectors of observed and expected counts at time j; there are m elements in each vector, corresponding to entries for each of the m regions. The vector of observed counts is assumed to be approximately multivariate normal. This will require either a sufficiently large number of counts or some transformation to normality. Qiu and Hawkins (2003) have recently suggested an alternative, nonparametric procedure for situations where the distribution is not multivariate normal. The quantity n_t represents the number of time periods since the cusum was last reset to zero, and is discussed further down in the text.

The norm of $\mathbf{S}_t$, $\|\mathbf{S}_t\|$, is a scalar representing the multivariate distance of the cumulated differences from the target:

$$\|\mathbf{S}_t\| = \sqrt{\mathbf{S}_t' \Sigma^{-1} \mathbf{S}_t} \tag{10.12}$$

where Σ is the variance–covariance matrix associated with the m regions. The quantity monitored is

$$MC1_t = \max\{0, \|\mathbf{S}_t\| - k n_t\} \tag{10.13}$$

where

$$n_t = n_{t-1} + 1; \quad MC1_{t-1} > 0$$

$$= 1; \qquad \text{otherwise} \tag{10.14}$$

The value of the parameter k is chosen to be equal to one-half of the multivariate distance from the target vector to the hypothesized alternative vector. The parameter k may also be thought of as equal to one-half of the off-target multivariate distance one would like to quickly detect; such a choice is thought to minimize the time taken to detect such an off-target process.

10.3.3.2 *Hypothetical, Simulated Scenarios*

To compare and illustrate the methods, we first simulated a 10-region system, where the baseline disease rate was 3.19 cases per thousand population. We took each region to have a population of 10,000 and assigned each person the disease with probability .00319. The distribution of disease counts is therefore binomial (and approximately normal), with mean 31.9, and variance $31.9 \times (1 - .00319)$. We then simulated cases for each region from this distribution and standardized to create a variable with mean zero and variance one. We looked at both the null hypothesis and alternatives where (1) one region was assigned a random number of cases that was one standard deviation above the mean of 31.9, and (2) three regions were each assigned a random number of cases that was one standard deviation above the mean of 31.9.

When the 10 regions were monitored separately, we used the values of h and k shown in Table 10.4, under the heading "univariate cusum." The

TABLE 10.4

ARL$_1$'s Associated with Simulated Alternative Hypotheses in a 10-Region System

(a) Increase of One Standard Deviation in One Region

	Univariate Cusum		Multivariate Cusum	
k	h	ARL$_1$	h	ARL$_1$
0.4	6.07	10.55	9.8	11.15
0.5	5.06	10.05	8.5	10.49
0.6	4.33	10.19	7.4	10.52
3.1	0	39.62	—	—
4.82	—	—	0	50.62

(b) Increase of One Standard Deviation in Each of Three Regions

	Univariate Cusum		Multivariate Cusum	
k	h	ARL$_1$	h	ARL$_1$
0.5	5.06	6.53	8.5	5.79
0.75	3.52	5.98	6.3	5.07
0.865	3.06	5.83	5.6	4.69
1.0	2.64	6.51	5	4.82
3.1	0	16.45	—	—
4.82	—	—	0	18.02

Note: Results are based upon 1,000 Monte Carlo simulations.

values of h were determined for each k value by using Equation 7.5 with $ARL_0 = 100 \times 10 = 1000$ (except for the case where $h = 0$; in this special case, the null hypothesis was simulated to determine that a value of $k = 3.1$ would yield a value of $h = 0$ for the desired ARL_0). False alarms therefore occur, on average every 100 observations, in at least one of the 10 regions. The case $h = 0$ always corresponds to monitoring the extreme values in each region; that is, this method is equivalent to the Shewhart chart discussed in Chapter 7, Section 7.2. The cusum will signal as soon as the threshold of $h = 0$ is exceeded, and this in turn will occur for the first observation that is greater than k. Thus, observations are evaluated individually until a region achieves a critical z-score of k or greater. Under the heading "multivariate cusum" is the critical value of h corresponding to $ARL_0 = 100$ and the value of k in the first column (where the value of h that is associated with k was determined by simulating the null hypothesis as Equation 7.5 does not apply to the multivariate case), as well as the ARL$_1$ resulting from application of Pignatiello and Runger's (1990) multivariate cusum method to 1,000 simulations of the alternative hypotheses. Panels (a) and (b) of Table 10.4 correspond to the alternative hypotheses (a) and (b), respectively.

For panel (a), the empirical optimal value of k for both univariate and multivariate cusums is 0.5; this value not only minimizes the time needed to detect the change but it also corresponds to the theoretical optimal found as one-half of the distance between the null and alternative vectors. In panel (b), the theoretical optimal value of k is again found by taking one-half of the distance between the vector under the null hypothesis ($z = \{0\ 0\ 0\ 0\ \ldots\ 0\}$), and the vector under the alternative hypothesis ($z = \{1\ 1\ 1\ 0\ 0\ \ldots\ 0\}$), giving rise to $k = \sqrt{3}/2 = 0.865$. This value of k also corresponds to the minimum ARL_1 found in each column.

Note that quicker detection is possible by using the cusum with an optimal value of k, instead of monitoring the values in each region with Shewhart charts (i.e., with $h = 0$). For the univariate cusum case, in panel (a), change is detected after about 10 periods instead of almost 40 periods; in panel (b), change is detected after about 6 periods instead of about 16. Similar results are seen in the multivariate cusum results. Note also that the univariate approach appears to be just slightly better than the multivariate method in panel (a), whereas the multivariate method is better in panel (b). We will discuss this finding in more detail after presenting the 200-region case.

We next simulated a 200-region system with alternative hypotheses (a) increasing the average number of cases in one region by one standard deviation, and (b) increasing the average number of cases by one standard deviation in each of five regions. Table 10.5 displays the results. As in Table 10.4, improvement is possible over either univariate or multivariate Shewhart charts by using a value of h other than zero. In panel (a) of Table 10.5, the univariate scheme is seen to perform significantly better than the multivariate scheme, as characterized by the lower ARL_1's. The minimum ARL_1 for the case of a change in one region is achieved at the optimal value of $k = 0.5$ in the univariate case; in the multivariate case, the minimum ARL_1 is achieved at a value of $k = 1.2$, which is higher than the expected optimal value of $k = 0.5$.

In panel (b), the optimal value of k is equal to $\sqrt{5}/2 = 1.12$ At this value of k, the ARL_1 associated with the univariate cusum is lower than that of the multivariate cusum. However, the ARL_1 for the univariate case can be lowered further by using lower values of k. Thus, the optimal choice of k, when the direction of change is not in the direction of a single variable, is no longer equal to one-half the distance between the null and alternative hypotheses. In addition, the ARL_1 for the multivariate case can be lowered further by using higher values of k. If the choice of k can be made well for the multivariate cusum (i.e., k is chosen to be about 2.4), the ARL_1 will be lower than that associated with the univariate case. We have found that, when the number of regions exceeds about 10 or 15, the choice of k equal to one-half of the distance to the alternative is no longer optimal for the multivariate cusum and, instead, larger values of k should be used. This finding deserves further study.

Note that in panel (b) of Table 10.5, the ARL_1's are again generally lower when individual, univariate cusums are employed. This is because the direction of the change in both (a) and (b) is significant enough in the direction

TABLE 10.5

ARL_1's Associated with Simulated Alternative Hypotheses in a 200-Region System

(a) Increase of One Standard Deviation in One Region

	Univariate Cusums		Multivariate Cusum	
k	h	ARL_1	h	ARL_1
0.4	9.5	16.31	96	61.5
0.5	8.045	16.2	87	55.5
0.7	5.9	16.5	68	40.9
1	4.15	24.8	50.5	28.8
1.2	3.45	30.1	43.1	24.8
1.4	2.9	40.4	37.7	26.3
3.89	0	84.3	14.7	79.2
15.8	—	—	0	88.9

(b) Increase of One Standard Deviation in Each of Five Regions

	Univariate Cusum		Multivariate Cusum	
k	h	ARL_1	h	ARL_1
0.5	8.045	9.5	87	31.9
0.7	5.9	8.49	68	24.0
0.8	5.2	8.6	61.5	21.6
0.9	4.6	8.56	55.7	18.8
1.0	4.17	9.1	50.5	16.9
1.12	3.7	10.1	46	15.0
2.2	1.75	45.0	25	7.95
2.4	1.5	45.3	23.1	7.63
2.6	1.3	49.7	21.45	8.26
3.89	0	52.5	14.7	21.8
15.8	—	—	0	55.6

of individual variables that the univariate approach can better detect it. Table 10.6 shows two other types of change of the same multivariate magnitude as in Table 10.5(b). The first type of change consists of an increase of 0.158 in the z-score of each region (corresponding to a distance between the alternative and the null of $\sqrt{\sum_{i=1}^{200} .158^2} = \sqrt{5}$), and the second consists of an increase of $\sqrt{5}$ in just one region. For the latter type of change (Scenario 1 in Table 10.6), the univariate cusum detects the change very quickly, and the optimal k occurs when it is equal to one-half the distance ($\sqrt{5}/2 = 1.12$) from the null to the alternative. When each region changes by a small amount as in the former type of change (Scenario 3), the univariate cusum does not do as well in detecting the change (and in fact the ARL_1's are longer than the corresponding ARL_1's for the multivariate case, also given in Table 10.6). In addition, the usual choice of k is not optimal.

TABLE 10.6

ARL_1's for Simulated Alternative Hypotheses with the Same Amount of Total Change in a 200-Region System

	Univariate Cusum			Multivariate Cusum		
	ARL_1			ARL_1		
k	Scenario 1	Scenario 2	Scenario 3	Scenario 1	Scenario 2	Scenario 3
0.5	5.2	9.5	23.6	31.93	31.9	31.88
0.7	4.5	8.49	25.3	23.9	24	23.89
0.8	4.3	8.6	26	21.64	21.59	21.26
0.9	4.2	8.56	28.4	19.08	18.84	18.97
1	4.1	9.1	31.8	17	16.89	16.96
1.12	4	10.1	32.1	15.16	15.04	15.12
1.3	4	—	—	—	—	—
2.2	—	—	45	—	—	—
2.4	—	—	45.3	7.68	7.75	7.64
2.6	—	—	49.7	—	—	—
3.89	—	—	52.5	20.18	21.83	19.57
5	—	—	—	34.38	33.38	32.24
15.8	—	—	—	51.11	55.57	54.86

Note: Scenario 1: increase of $\sqrt{5}$ standard deviations in one region; Scenario 2: increase of one standard deviation in each of five regions; Scenario 3: increase of 0.158 standard deviations in 200 regions.

As Pignatiello and Runger (1990) note, the individual cusum approach is *directional*; the ARL_1 will differ for different types of change of the same multivariate magnitude. The ARL_1 will be very low when the change is concentrated in a single region, and will be very high when the same total amount of change is spread across many regions. In contrast, the multivariate cusum method is *nondirectional*; its ARL_1 properties are the same for a change of given multivariate magnitude, regardless of the direction of change. Table 10.6 demonstrates this; for any given combination of k and h that is used, the ARL_1 associated with the multivariate cusum for the three scenarios (which represent three different types of change, all of the same magnitude) remains practically the same.

10.3.3.3 Spatial Autocorrelation

The previous illustration and results have assumed that there is no spatial autocorrelation in the underlying regional observations. In many situations, this will be unrealistic, and this is especially likely if the spatial units are small.

To investigate the effects of spatial autocorrelation, we first simulated a simultaneous autoregressive model for a 10-region configuration. All regions are square cells; nine of them are arranged into a three-by-three grid, and the tenth is located below it in the first column. We assumed rook's case adjacency and carried out a multivariate cusum approach to surveillance.

Figure 10.1 summarizes the results of simulations of alternative hypotheses for various degrees of spatial autocorrelation for various values of k (for each k there is an associated value of h that is consistent with $ARL_0 = 100$). The results also reflect the assumption that the degree of spatial autocorrelation is known, and is captured in the variance–covariance matrix Σ. Panel (a) shows that, when the mean in one of the regions (in this case, region 1) increases by one standard deviation, the time to detection declines with increasing spatial autocorrelation. This is because the multivariate distance between the null and alternative hypotheses is increasing with increasing spatial autocorrelation, as measured by ρ (see Figure 10.2). It also implies that the optimal value of k increases with increasing ρ.

Panel (b) displays the results when there is an increase of one standard deviation in the mean values of each of three regions (regions 1, 2, and 3). The time to detection is slightly higher for the cases $\rho = 0.2$ and 0.5; note from Figure 10.2 that the multivariate distance (and hence, optimal choice of k) is slightly lower in these two cases, in comparison with the cases where $\rho = 0.0$ and 0.7.

Because the degree of spatial dependence influences the multivariate distance between the null hypothesis and specified changes in subsets of regions, the ARL_1 is affected as well. Panel (c) of Figure 10.1 depicts simulations where there is a change of $\sqrt{1/3}$ in each of three regions (regions 1, 2, and 3). When there is no spatial autocorrelation, ARL_1's are the same as in panel (a) because the multivariate distances between the null and the alternative are the same in these two cases. A comparison of the $\rho = 0.7$ lines of panels (a) and (c) reveals that it takes much longer to detect small changes spread across a number of regions because these regions' observations are spatially autocorrelated. This is captured in Figure 10.2, which shows that the multivariate distance for the $\rho = 0.7$ case is in fact much smaller in the panel (c) case than in the panel (a) case.

In addition to the effects of spatial autocorrelation just described, ARL_0's can be under- or overestimated because the covariance matrix that captures spatial dependence is typically assumed or estimated, and not known exactly. Figure 10.3 (a) shows that, when the covariance matrix is incorrectly taken to be the identity matrix (i.e., when it is incorrectly assumed that there is no spatial dependence), the presence of spatial autocorrelation leads to much lower ARL_0's in comparison with the nominal ARL_0 of 100 in this example. Thus, premature signals might be caused by spatial autocorrelation and may not represent true increases in incidence. Furthermore, the effect on ARL_0 is more serious when k is high than when it is low. This would suggest that it is better to choose a value of k that is less than optimal than choosing value that is greater than optimal. Note, also, that these effects are tempered somewhat by employing the multiple

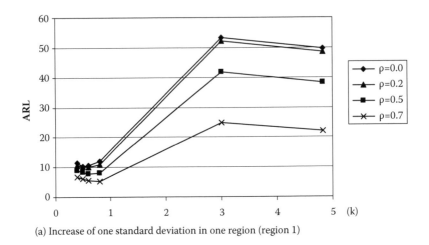

(a) Increase of one standard deviation in one region (region 1)

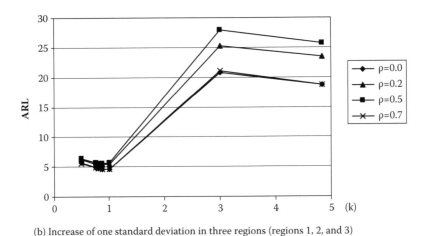

(b) Increase of one standard deviation in three regions (regions 1, 2, and 3)

FIGURE 10.1
ARL₁s under alternative hypotheses with varying degrees of spatial autocorrelation.

univariate cusum instead of the multivariate cusum. Panels (b) and (c) provide additional examples when estimated values of $\rho = 0.35$ and 0.6 are used, respectively. Note from these two panels that the effects of over-estimating ρ are less serious than the effects of underestimating it.

10.3.4   Example: Breast Cancer Mortality in the Northeastern United States

To illustrate, we use the annual data from the CMF on the observed and expected number of breast cancer cases in the 217 counties comprising the

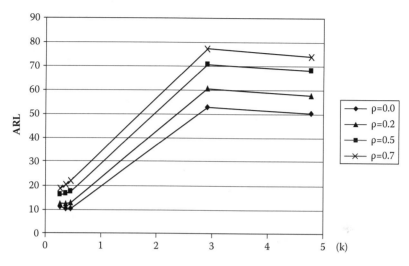

(c) Increase of $\sqrt{\frac{1}{3}}$ standard deviation in three regions (regions 1, 2, and 3)

FIGURE 10.1
(Continued)

northeastern United States. We used the first 10 years (1968–77) as a base period, finding the mean annual relative risk (O/E) for each county. We next adjusted county-specific expectations by multiplying expectations for all subsequent years by the mean annual relative risk observed for the county during the base period. Surveillance then began in the 11th year for which there is data (1978).

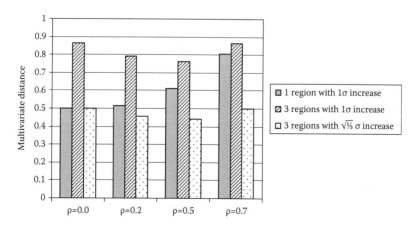

FIGURE 10.2
Multivariate distances associated with the three scenarios shown in Figure 10.1.

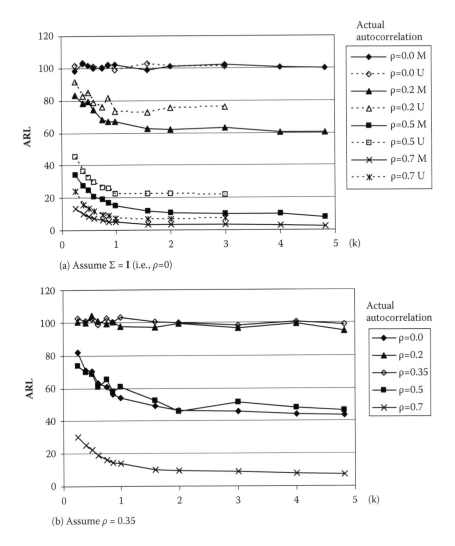

(a) Assume $\Sigma = \mathbf{I}$ (i.e., $\rho=0$)

(b) Assume $\rho = 0.35$

FIGURE 10.3
ARL_0s for varying degrees of actual and assumed spatial autocorrelation (M: multivariate cusum; U: multiple univariate cusum).

10.3.4.1 Multiple Univariate Results

To monitor all 217 counties simultaneously, with an average run length of 100 years for the first false alarm to occur in at least one county requires setting $ARL_0 = 21{,}700$ in Equation 7.4 when $k = 1/2$ is used. Doing so results in a value of $h = 8.13$. Table 10.7 displays the results. The first alarm is sounded in the 21st year for which there is data (1988), in Philadelphia

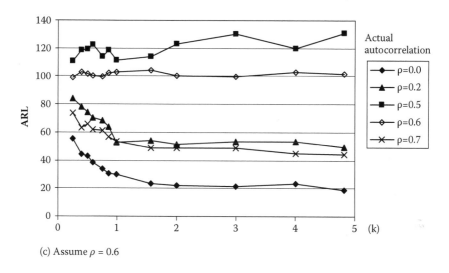

(c) Assume $\rho = 0.6$

FIGURE 10.3
(Continued).

County, Pennsylvania. The cusum for this county is shown in Figure 10.4; it increases steadily after about 1983 or 1984, and first exceeds the critical value of $h = 8.13$ in 1988. The ratio of observed to expected breast cancer cases is shown for this county in Figure 10.5. In 4 of the 6 years up to and including 1988, the ratio of observed to expected cases exceeded 1.0. Prior to the period 1983–88, the observed/expected ratio exceeded 1.0 in only four of the previous 15 years. In 1989, Delaware County, Pennsylvania,

TABLE 10.7

Results of the Multiple Univariate Cusum Analysis of Breast Cancer Mortality in the Northeastern United States

County	First Signal	Signaling Period
Berkshire, MA	28	28–31
Hampshire, MA	31	31
Belknap, NH	24	24–26, 28–29, 31
Coos, NH	25	25–31
Grafton, NH	24	24–27
Cape May, NJ	24	24–26, 31
Delaware, PA	22	22–23, 25, 27, 29, 30
Huntingdon, PA	25	25–26, 28
Juniata, PA	27	27, 31
Montour, PA	26	26–31
Philadelphia, PA	21	21–31
Addison, VT	30	30–31

Note: Year 20 = 1987; critical threshold: $h = 8.13$; $k = 0.5$.

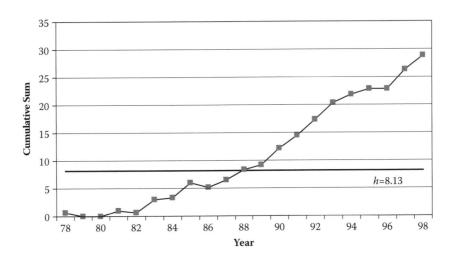

FIGURE 10.4
Cusum for Philadelphia County, 1978–1998.

also signals a change from the base period. Note that, for some counties, the signal is persistent (e.g., Philadelphia), whereas for others it is not (e.g., Juniata, Pennsylvania, where the cumulative sum exceeds its critical value only in 1994 and 1998). The results are broadly consistent with those of Han and Rogerson (2003); they found an increasing amount of geographic clustering in the Philadelphia region during the 1990s.

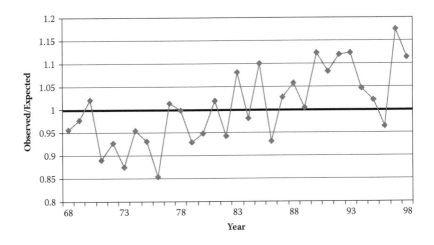

FIGURE 10.5
Ratio of observed to expected breast cancer cases for Philadelphia County, 1968–1998.

10.3.4.2 Multivariate Results

If we wish to minimize the time to detect a change of one standard deviation in the rate of any of the 217 counties (or the equivalent multivariate change when the change is spread out among more than one county), we choose $k = \frac{1}{2}$, which is one-half of the magnitude of the change we wish to detect. By simulating the null hypothesis with $k = \frac{1}{2}$, we found that $h = 93$ yields an average run length of 100. We also assumed a variance–covariance matrix equal to the identity matrix; this decision was based upon the fact that the value of Moran's I, a measure of spatial autocorrelation, was significant (using $\alpha = 0.05$) in only one of the first 10 years.

When these parameters are then used to monitor the observed counts beginning in 1978 in relation to the expected counts described in the previous subsection, a signal is first sent in 1989 (1 year after the multiple univariate signal was sent). Figure 10.6 shows how the MC1 values change over time; the univariate cusum for Philadelphia is superimposed on this figure to show how it helps drive the multivariate cusum.

When that is not the case, the multivariate approach is generally preferred. The multivariate approach seems to be generally preferred when change is not confined to a small number of regions. We now turn to the interpretation of the multivariate signal.

10.3.4.3 Interpretation of Multivariate Results

There have been numerous suggestions made for interpreting multivariate charts when they signal. For example, Pignatiello and Runger (1990) suggest that "looking at individual measurements along with the principal components can provide an insight." Jackson (1985) suggests using

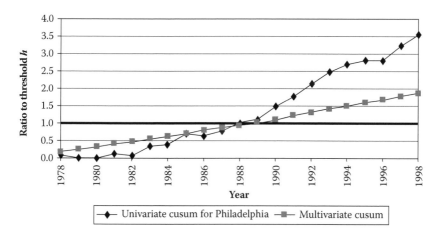

FIGURE 10.6
MC1 values over time for breast cancer mortality data, 1978–1998.

information on the principal components to preserve information on the correlation between variables. Lowry and Montgomery (1995) reinforce this by noting that "the combination of using a multivariate control chart for signaling purposes and then using separate charts for diagnostic purposes is often effective." Alt (1984) recommends using Bonferroni limits on individual variables. The approach we use here is similar in spirit to that of Alt.

When the multivariate cusum signals, the quantity $||\mathbf{S}_t||$ may be decomposed to examine the contribution from individual regions. In particular,

$$||S_t|| = \sum_i z_{it}^2 \qquad (10.15)$$

where

$$z_{it} = \sum_{j=t-n_t+1}^{t} \frac{O_{ij} - E_{ij}}{\sqrt{\lambda(1-\lambda)P_{ij}}} \qquad (10.16)$$

and where λ is the rate of disease, and P_{ij}, O_{ij}, and E_{ij} are, respectively, the population, observed count, and expected count for region i at time j. Note that O_{ij} and E_{ij} are elements of vectors O_j and E_j defined previously. Dividing this by $\sqrt{n_t}$ yields a quantity that, for a given region i, will have a standard normal distribution. Because we are testing the significance of z_i in m separate regions, we need to evaluate whether z_i is greater than a critical value determined using the Bonferroni correction, implemented by using α/m instead of α in the tail of the distribution. In our case, $m = 217$, and setting $\alpha = 0.05$ implies that the critical value of z is $\Phi^{-1}(1-\alpha/m) = 3.50$. Because the cusum signals in the 12th year of surveillance, $n_t = 12$, and for a region to contribute significantly to the signal we require $z/\sqrt{12} > 3.5$, or $z > 12.13$. Table 10.8 reveals that the counties that contribute

TABLE 10.8

Regions with Large Values of z_i in 1989, Contributing to the Multivariate Signal

County	z_i
Philadelphia, PA	14.82
Delaware, PA	13.49
Ontario, NY	13.40
Cape May, NJ	13.38
Berkshire, MA	13.00
Belknap, NH	12.39
Schuylkill, PA	12.24

to the signal are similar to those found using the simultaneous univariate approach.

10.3.4.4 *Estimation of Covariance and a Nonparametric Approach*

In comparison with multiple univariate surveillance, multivariate surveillance has the advantage of better detecting small changes that occur in a sizable number of regions. This is particularly true when there is significant underlying covariation in regional values because the univariate approach does not take the covariation into account. However, parametric multivariate surveillance in a regional context requires estimation of the regional variance–covariance matrix. When vectors of regional observations are available for a large number of time periods, it is possible to estimate the covariance matrix directly. For m regions, one would like to have at least m time periods, and preferably many more. In this regard, Daniels and Kass (2001) discuss estimators that shrink the eigenvalues of the matrix because the largest eigenvalue tends to be too large, and the smallest too small.

Another alternative would be to estimate a spatial model, such as the simultaneous autoregressive model used in the simulations of the previous section. This has the advantage of parameterizing the spatial structure (as opposed to estimating a much larger number of elements of the covariance matrix directly), but is typically based on data from a single time period, and so the assumption must be made that the spatial structure estimated from that one period will remain constant over time. Finally, as mentioned earlier, it is possible to adopt a nonparametric approach that eliminates the need to make assumptions about multivariate normality. In the remainder of this subsection, we report on the results obtained when using these alternatives on a subset of the breast cancer data that contains the 10 counties of New Hampshire.

Table 10.9 summarizes the results from the multivariate monitoring method described in subsection 10.3.3.1, when using (a) the identity matrix as the covariance matrix, where no regional covariation is assumed, (b) a covariance estimated using the simultaneous autoregressive model, with parameters taken as the average of the annual parameters observed in each of the first 10 years of data, and (c) using the log eigenvalue shrunken estimator described by Daniels and Kass (2001). For the latter, we separately used the first 10, 15, and 20 years of data to estimate the covariance matrix because the results using only 10 year's data yielded a variance–covariance matrix with determinant close to zero. In all cases, we used a critical threshold h that was consistent with an ARL_0 of 100. In interpreting the results, it should be kept in mind that the purpose of this is to simply compare the similarity and differences of the approaches; we cannot declare any single approach "better" or "best" because we do not know the true parameters. For example, signaling earlier is not necessarily "better" because it is not known whether true change has occurred. The

TABLE 10.9

Comparison of Various Cusum Methods Applied to Breast Cancer Mortality Data for Northeastern United States

Method		Spatial Structure	First signal
	Univariate cusum	(Not available)	1985 (Grafton, NH)
		Assuming no spatial covariation	1984
Parametric		Using simultaneous autoregressive model	1982
	Multivariate cusum	Shrinking covariance matrix estimated from 10 years' data	1978
		Shrinking covariance matrix estimated from 15 years' data	1983
		Shrinking covariance matrix estimated from 20 years' data	1984
Nonparametric	Multivariate cusum	(Not available)	1991*

* The nonparametric multivariate cusum did not reach its threshold; it came closest to the critical threshold in 1991.

comparison remains useful in illustrating both the variety of approaches to treating distributional assumptions and covariance estimation, and the similarities and differences in the results that are possible.

Multivariate surveillance using the identity matrix as the covariance matrix leads to a signal in 1984. When using a 10-year average of the annual simultaneous autoregressive model parameters (where the autoregressive parameter ρ is estimated to be 0.14), the first signal occurs in 1982. A check of the individual univariate cusums shows Grafton, Belknap, and Coos counties all contributing to this signal.

Simply using the first 10 years' data to estimate the variance–covariance matrix directly results in a noninvertible matrix; shrinkage yields an invertible matrix, but it is quite unstable (the cusum signals in the first year, with a value of 57, and rises to 547 by 1998). Using the first 15 years' data together with a shrunken estimator results in a multivariate cusum which signals in 1983 and, after 1984, rises sharply in comparison with the other approaches. This again can be most likely attributed to instability in the estimated matrix. With 20 time periods used in the estimation the cusum signals in 1984 are more similar to the results from the identity matrix and the spatially autoregressive model. This underscores the necessity of having a sufficient number of time periods when estimating the variance–covariance either directly or using shrunken estimators.

Daniels and Kass (2001) do suggest additional shrinking toward some underlying structure, but such shrinkage toward an underlying spatial structure has not been attempted here.

We also used the nonparametric approach recently suggested by Qiu and Hawkins (2003). This method is based on the frequency with which a given region has the maximum value among the values observed in all m regions. A description of a simplified version of the method is as follows. The cusum is

$$y_n = (\mathbf{S}_n^{(1)} - \mathbf{S}_n^{(2)})' diag(1/S_{n,1}^{(2)}, \ldots, 1/S_{n,m+1}^{(2)})(\mathbf{S}_n^{(1)} - \mathbf{S}_n^{(2)}) \qquad (10.17)$$

where $\mathbf{S}_n^{(1)}$ and $\mathbf{S}_n^{(2)}$ are $(m+1)$-vectors with element i defined, respectively, as the observed and expected counts of the event (across all time periods, up to and including n) when region i has the maximum value among all regions. The threshold value h is determined by simulating the null hypothesis that all regions have an equal chance of being the region with the largest observed value in any one time period. We implemented their method (together with an "allowance constant" of 0.5 and accounting for the possibility that change may occur simultaneously in all regions; for details see Qiu and Hawkins 2003) and found the critical threshold to be 12.5 (consistent with an ARL_0 value of 100). The maximum cusum was 11.88, in 1991; in this year Belknap had the highest standardized rate of breast cancer among the 10 counties—and this was the fourth time this had occurred since monitoring began in 1978. During 1978–1991, Coos County also had the highest standardized rate in four separate years.

10.3.5 Discussion

The univariate approaches are limited by their lack of ability to account for the spatial autocorrelation of regional data; the multivariate methods are limited by the difficulty in accurately specifying the multiregional covariance structure. When the degree of spatial autocorrelation is low, the univariate method is generally better at detecting changes in rates that occur in a small number of regions; the multivariate is better when change occurs in a large number of regions.

In general, when the degree of spatial autocorrelation is strong and can be well estimated, multivariate monitoring is preferred because the multiple univariate approach has no way of handling spatial autocorrelation. Although multivariate monitoring has a strong conceptual appeal due to its generality, in practice it does not necessarily have an advantage over multiple univariate cusums, especially when the effects of spatial autocorrelation are weak or are poorly estimated. Multiple univariate cusums are particularly well suited to the situation where there is little or no spatial autocorrelation, and where changes are anticipated in a relatively small number of regions.

11

Summary:
Associated Tests for Cluster
Detection and Surveillance

11.1   Introduction

The primary purpose of this final chapter is to present a simplified and unified approach to retrospective and prospective detection of spatial clustering. In doing so, we wish not only to convey the relationships among the alternative tests but also to provide a relatively straightforward description of the individual tests to facilitate their implementation. The *GeoSurveillance* software is also structured in much the same way that this chapter is organized. However, as noted at the beginning of the book, the reader should be aware that this is just one of several alternative approaches to the subject.

In Chapter 1, Section 1.5, Besag and Newell's three-way classification of statistical tests for clustering was introduced. Chapters 3, 4, and 5 were devoted to discussions of each of these three types. Although there are many approaches used for each of these types, they have generally been developed independently of one another and are not conceptually integrated. Some local and global statistical measures have been integrated (e.g., the global and local Moran statistics discussed in Chapter 3, Section 3.4, and Chapter 4, Section 4.2, respectively); in fact, a desirable property of local statistics is that they add to a multiple of the global statistic (Anselin 1995).

In Section 11.2, we shall focus more closely on the connections and relationships between the three types of cluster detection tests. In Section 11.3, we recap some of the material in Chapter 10, indicating how the associated retrospective tests in Section 11.2 can be generalized for use in a surveillance (i.e., prospective) setting.

The classification outline in Table 11.1 is not meant to be exhaustive; instead, its purpose is to provide a typology within which the associated methods described in this chapter fit. The resulting organization structure is also useful in understanding the structure of the *GeoSurveillance* software.

TABLE 11.1

Classification of Associated Retrospective and Prospective Tests
for Cluster Detection

	Aspatial Methods	Spatial Methods
I. Associated retrospective tests (Section 11.2)		
Global/general	Chi-square goodness-of-fit	Spatial chi-square statistic Global score statistic
Local/focused	Poisson test Binomial test	Score test
Detection of clustering (maximum of local tests)	*M*-test	Spatial *M*-test
II. Prospective tests for monitoring and quick detection of new clusters (Section 11.3)		
	Univariate cusum charts— one for each spatial unit	Univariate cusum charts—one for each spatial unit and its surrounding neighborhood
	Bonferroni adjustment	Adjustment for testing of multiple neighborhoods

11.2   Associated Retrospective Statistical Tests

The complementarity of associated statistical tests was noted earlier in an aspatial context by Fuchs and Kenett (1980); they derived the *M*-test based on the largest outlier in a multinomial distribution (see Chapter 4, Section 4.9). They noted that, when the common chi-square statistic (a global statistic) comparing observed and expected frequencies across categories is rejected, the *M*-test may be used to determine *which* cells were most responsible for the rejection of the null hypothesis. Their *M*-statistic is equivalent to the maximum local (aspatial) statistic and is based on the cell or category that contributes most to the global chi-square statistic. In addition, the *M*-test is shown to have relatively higher power than the global statistic under the alternative hypothesis that there are a small number of outlying cells. One may therefore find individual cells that are inconsistent with the null hypothesis even when the global chi-square statistic is insignificant. The relatively higher power of the *M*-test against an outlier alternative in comparison with the global chi-square test is because the *M*-test is designed specifically for, and is based on, outlying values from the multinomial distribution. This is an example of how related tests can be used in conjunction with one another.

The purpose of the subsections 11.2.1 and 11.2.2, which are based upon Rogerson (2005a), is to describe a set of associated statistical tests for both aspatial and spatial clustering. In subsection 11.2.1, associated tests are described for the aspatial case, and in subsection 11.2.2, the details of the associated set of statistical tests for the spatial case are described.

11.2.1 Associated Retrospective Statistical Tests: Aspatial Case

A focused or local test for a particular region can be based on

$$z_i = \frac{O_i - E_i}{\sqrt{E_i}} \tag{11.1}$$

where O_i and E_i refer to the observed and expected counts for region i. (We assume that the expected counts refer to an expectation under the null hypothesis of no raised incidence in region i. In the simplest case, these expectations are found by multiplying the size of the at-risk population in region i by the common, overall disease rate. In more complex examples, the expected number of cases in region i could be found as the output of a separate model that predicted the number of cases in region i as a function of not only population but also other relevant covariates such as age, gender, income, education, etc.)

The quantity in Equation 11.1 will have, approximately, a standard normal distribution under the null hypothesis of no raised incidence in region i. This is based on the normal approximation to the Poisson distribution (and certainly the validity of this approximation needs to be considered when expectations are small; expectations as low as one typically give acceptable results). This local statistic can then be assessed by comparing the observed z value with tabled values of the normal distribution.

It is also possible to use z_i^2 as the local statistic; in this case, the statistic will have a chi-square distribution with one degree of freedom under the null hypothesis of no raised incidence in that region. (A disadvantage of this form is that, because the z value is squared, it does not distinguish between positive and negative deviations from expectations.) The sum of the local statistics expressed in this form is equal to the global statistic, which is a desirable property of local statistics as mentioned earlier. Specifically, the global statistic is the familiar chi-square goodness-of-fit statistic:

$$\chi^2 = \sum_{i=1}^{m} \frac{(O_i - E_i)^2}{E_i} \tag{11.2}$$

With m regions, this global statistic has $m - 1$ degrees of freedom.

Finally, it is of interest to ask how we can assess the entire set of local statistics simultaneously. After all, it is more common that we will not know where to look for raised incidence; local statistics allow us to look at any one prespecified region, but we must account for multiple testing if we examine all regional local statistics. Fuchs and Kennett (1980) suggest that the maximum local statistic (i.e., $\max_i z_i$) be assessed by using the normal distribution with a critical value based on α/m instead of the usual value of α, where α is the probability of a type I error, that is, the probability that a true null hypothesis is rejected. This implementation of the M-test amounts to a Bonferroni adjustment for the number of tests (m) being carried out. In actuality, the z-tests for each region are negatively correlated with one another; if a high count is observed in one region, this implies a low count in some other region because the total number of cases is fixed. These negative correlations may be used in the derivation of a lower bound for the critical value (see Fuchs and Kenett 1980); the critical value of the test statistic based on an upper tail of α/m provides a conservative, upper bound for the critical value.

For the scale on which the data have been collected, the common chi-square goodness-of-fit test, the local test, and the M-test provide an associated set of tests allowing for the examination of clustering of all three types, namely, general tests, focused tests, and tests for the detection of clustering. The example is aspatial because the local statistic for each region is based on data for that region only. Generalization to facilitate detection of clusters on larger spatial scales is treated in the following subsection.

11.2.2   Associated Retrospective Statistical Tests: Spatial Case

More generally, we would often like to base local statistics on data from not only the region of interest but from surrounding regions as well. This section uses the structure for the aspatial tests outlined in the previous section, extending it to the case in which local statistics are based on data from a neighborhood surrounding the region of interest.

The set of tests described in subsection 11.2.1 will prove most effective in detecting deviations from the null hypothesis when the subarea with raised incidence occurs at the spatial scale of the regional unit. If, however, there is a spatial cluster of regions that has raised incidence (in other words, an increase in incidence has occurred over a set of clustered regions rather than in a single region) more explicitly, spatial methods are called for.

One way to incorporate such spatial elements more explicitly is to use local statistics is based not only on the target region but also on its neighbors. However, local statistics are correlated in space; this complicates the statistical assessment of the maximum local statistic. Most tests, such as Kulldorff's spatial scan statistic, rely on Monte Carlo simulation to assess significance of many correlated local cluster tests. The tests summarized here are advantageous in that they do not require Monte Carlo simulation.

We begin this subsection with a review of the score statistic (e.g., see Lawson 1993; Waller and Gotway 2004), a special case of which will serve as our local statistic for focused tests. Recall from Chapter 4, Section 4.3 that, to test for raised incidence around region i, the score test is based on the following statistic:

$$U_i = \sum_{j=1}^{m} w_{ij}(O_j - E_j) \qquad (11.3)$$

where w_{ij} is the weight associated with the relationship between regions i and j. This is an alternative form of Equation 4.4 (Chapter 4), and under the null hypothesis of no clustering around location i, it has a normal distribution with mean zero and variance:

$$V[U_i] = \sum_j w_{ij}^2 E_j - \left(\sum_j O_j \right) \left(\sum_j w_{ij} \left\{ E_j / \sum_j E_j \right\} \right)^2 \tag{11.4}$$

which is an alternative form of Equation 4.5 (Chapter 4).

In the special case in which the weights are defined using a scaled Gaussian kernel and then divided by the square root of the expectations, $V[U_i] = 1$, and the local statistics can be taken as coming from a standard normal distribution under the null hypothesis of no raised incidence in the vicinity of the location.

In particular, the Gaussian weights are defined as follows:

$$w_{ij} = \frac{\sqrt{A}}{\sqrt{m\pi}\sigma} \exp\left(-\frac{d_{ij}^2}{2\sigma^2(m/A)} \right); \quad i,j = 1, 2, \ldots, m \tag{11.5}$$

where σ is the bandwidth defining the local neighborhood, and d_{ij} is the distance from region i to region j (e.g., the distance between regional centroids). A is the area of the study region, and m is the number of subregions. With this definition, $\sigma = 1$ implies that the standard deviation of the Gaussian kernel extends from the center of the region to a distance equal to the average distance between regional centroids. As σ increases, more smoothing occurs, and weights given to distant subregions increase.

These weights are also scaled, so that approximately, $\sum_j w_{ij}^2 = 1$. This approximation breaks down somewhat for subregions that are heterogeneous in size and shape, and also breaks down somewhat near the edges of the study region. The property can be restored by adjusting the weights as follows:

$$\tilde{w}_{ij} = \frac{w_{ij}}{\sqrt{\sum_j^m w_{ij}^2}} \tag{11.6}$$

With these weights, the local statistic

$$U_i = \sum_j^m \left(\frac{\tilde{w}_{ij}}{\sqrt{E_j}} \right)(O_j - E_j) \tag{11.7}$$

is a weighted sum of regional z-scores (based on the Poisson distribution; see Equation 11.1) that are in the neighborhood of region i. Again, this local statistic has a standard normal distribution under the null hypothesis when the weights are defined as earlier. Also, the local statistic U_i^2 has a chi-square distribution with one degree of freedom. As the bandwidth approaches zero, the local statistics approach the aspatial local statistics defined in Equation 11.1.

Of course, an alternative local statistic, unadjusted for expectations, is simply

$$U_i = \sum_j^n w_{ij}(O_j - E_j) \tag{11.8}$$

which is approximately normally distributed, with a mean of zero and a variance given by Equation 11.4. Alternatively, using Equation 11.8, the statistic $U_i^2/V[U_i]$ has a chi-square distribution with one degree of freedom. The test is also known as the *Poisson trend test* and is a locally most powerful test (Breslow and Day 1987). Waller and Lawson (1995) show it to be powerful in rejecting false null hypotheses, in comparison with other tests.

In general, one should choose the bandwidth (σ) to match the anticipated cluster size; this will maximize the statistical power of detecting an actual cluster of that size. The matched filter theorem (Rosenfeld and Kak 1982) implies that, to maximize the statistical power of cluster detection, the size and shape of the chosen kernel should match that of the actual cluster. Thus, the choice of a Gaussian kernel is somewhat limiting; it will be a good choice if the actual cluster has risk that declines according to a Gaussian function, but will not be as good a choice if, for example, the alternative turns out to be a hot spot, where risk is uniformly elevated throughout the hot-spot region. In the latter case, a rectangular kernel would provide greater power. Siegmund and Worsley (1995) do find that Gaussian kernels are better at finding rectangular, hot-spot clusters than rectangular kernels are at finding Gaussian clusters, and this provides some consolation for the relative robustness of the Gaussian kernel. In addition, the literature on kernel density estimation often emphasizes that the size of the kernel chosen is more important than the shape of the kernel.

The corresponding global statistic U^2 is the sum of the squares of these local statistics. These can be written as

$$U^2 = \sum_i \sum_k \frac{w_{ki}^2(O_i - E_i)^2}{E_i} + \sum_i \sum_{j \neq i} \sum_k \frac{w_{ki}w_{kj}(O_i - E_i)(O_j - E_j)}{\sqrt{E_i E_j}} \tag{11.9}$$

(corresponding to the local statistics defined in Equation 11.7), and

$$U^2 = \sum_i \sum_k w_{ki}^2 (O_i - E_i)^2 + \sum_i \sum_{j \neq i} \sum_k w_{ki} w_{kj} (O_i - E_i)(O_j - E_j) \quad (11.10)$$

(corresponding to the local statistics defined in Equation 11.8).

The global score statistic defined in Equation 11.9 approaches the chi-square goodness-of-fit statistic for small values of the bandwidth σ. It can be seen as a spatial chi-square statistic; it is a combination of an aspatial goodness-of-fit measure and a Moran-like expression for pairs of regions.

The results of Tango (1995) can be used to assess the statistical significance of U^2. (Note that, although Tango discusses a focused score statistic, his general test is not developed as the sum of score tests.)

In particular, to assess the significance of Equation 11.10, let

$$a_{ii} = \sum_k w_{ki}^2$$

(11.11)

$$a_{ij} = \sum_k w_{ki} w_{kj}$$

$$E[U^2] = \frac{1}{n} Tr(\mathbf{A} \mathbf{V_p})$$

(11.12)

$$V[U^2] = \frac{2}{n^2} Tr(\mathbf{A} \mathbf{V_p})^2$$

where n is the total number of cases, and

$$\mathbf{V_p} = \Delta(\mathbf{p}) - \mathbf{p} \mathbf{p}'$$

(11.13)

where $\mathbf{p}$ is an $m \times 1$ vector of the expected proportion of cases falling in each of the m regions, and $\Delta(\mathbf{p})$ is an $m \times m$ matrix with the elements of $\mathbf{p}$ along the diagonal and zeros elsewhere. Then,

$$z = \{U^2/n - E[U^2]\}/\sqrt{V[U^2]}$$

(11.14)

and a test of the null hypothesis may then be carried out by treating the quantity in Equation 11.14 as a standard normal variate.

A chi-square approximation will generally be more accurate than a normal approximation. Again, following Tango, the statistic $v + z\sqrt{2v}$ has a chi-square distribution with v degrees of freedom, where

$$v = \left(\frac{\left\{ Tr(\mathbf{A} \mathbf{V_p})^2 \right\}^{1.5}}{\left\{ Tr(\mathbf{A} \mathbf{V_p}) \right\}^3} \right)^2$$

(11.15)

Significance tests for the global statistic as defined in Equation 11.10 proceed in the same manner, but there the weights are first divided by the expectation terms before proceeding with Equations 11.11–11.15.

11.2.3   Maximum Local Statistic

The issue of multiple testing arises when one wishes to carry out more than one focused test. It would be of interest, for example, to carry out local, focused tests at each of m locations and then declare which of the tests were significant. Appropriate decision making in this context requires one to account for the fact that multiple hypotheses are being tested simultaneously. A Bonferroni correction, in which each local test is carried out using a Type I error probability of α/m instead of α, is conservative because the local tests carried out at nearby locations are correlated with one another.

Using the results from Chapter 5, Section 5.10, the maximum local statistic (i.e., max U_i, when U_i is based on Equation 11.7) will exceed the following critical value with probability α, under the null hypothesis (Rogerson 2000):

$$U_\alpha^* \approx \sqrt{-\sqrt{\pi}\ln\left(\frac{4\alpha(1+0.81\sigma^2)}{m}\right)} \qquad (11.16)$$

where σ is defined as earlier (i.e., σ has been standardized by dividing by the square root of the average area of a region, i.e., $\sqrt{A/m}$, where A is the total area of the region being studied). In this way, it is interpreted as the width of the Gaussian kernel, expressed in multiples of the square root of the average regional area. Equivalently, the maximum squared score statistic, $U_{max}^2 = \max_{i=1,\dots,m} U_i^2$, will exceed its critical value of

$$U_{max,\alpha}^{2*} = -\sqrt{\pi}\ln\left(\frac{4\alpha(1+0.81\sigma^2)}{m}\right) \qquad (11.17)$$

with probability 2α, under the null hypothesis.

These critical values can also be used with the maximum among the U_i as defined in Equation 11.8 if the quantities are first standardized by dividing by the square root of the variance, given in Equation 11.4.

The set of three associated tests elaborated upon here (i.e., the global statistic U^2, the local statistic U_i^2, and U_{max}) has a number of distinctive and attractive features:

1. Any particular value of U_i^2 has a chi-squared distribution with one degree of freedom, and this can be used to test the null hypothesis of no raised incidence in the neighborhood of region i.

2. The probability that the maximum value of U_i^2 that is observed across all m regions will exceed $U^{2*}_{max,\alpha}$ with a given value is known. This is important because it allows for the simultaneous evaluation of all local statistics. It therefore avoids the need for either Monte Carlo simulation or a conservative Bonferroni adjustment.

3. The significance of the global statistic U^2 can be tested using critical values derived analytically.

4. The local statistics U_i^2, when summed across all regions, are equal to the global statistic. A significant global statistic will naturally lead to a search for those regions that contributed to the significance, and it can be assessed by searching for significant local statistics. If the global statistic is not significant, it is still possible that there exist regions with neighborhoods for which the z-scores are above what might be expected by chance.

It should also be pointed out that the local z-scores for each region can come from a Poisson specification but other alternatives are also possible, including the transformation suggested by Rossi et al. (1999):

a) $$z_j = \frac{O_j - E_j}{\sqrt{E_j}} \quad (Poisson)$$

b) $$z_j = \sqrt{O_j} + \sqrt{O_j + 1} - \sqrt{4E_j + 1} \quad (Freeman - Tukey)$$

c) $$z_j = \frac{O_j - 3E_j + 2\sqrt{O_j E_j}}{2\sqrt{E_j}} \quad (Rossi) \tag{11.18}$$

11.2.4   Illustration

Assume a study region consisting of nine regular grid cells ($m = 9$) and $\sigma = 1.2$. A Bonferroni-adjusted critical value associated with testing nine score statistics would be $\Phi^{-1} (1 - 0.05/9) = 2.54$ for $\alpha = 0.05$. Simulation of the null hypothesis (i.e., placing standard normal random variables in each of the nine grid cells) reveals that the 95th percentile of the maximum U value is 2.21, confirming the conservative nature of the Bonferroni adjustment. Equation 11.16 yields $U^*_{\alpha = .05} = 2.32$, which is only slightly higher than the value found by simulation.

11.2.5   Example: Application to Leukemia Data for Central New York State

In this section, we apply the methods described earlier to leukemia data analyzed by Waller et al. (1992, 1994) and Waller and Gotway (2004)

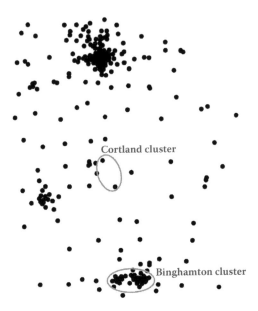

FIGURE 11.1
Census tract centroids in an eight-county region of central New York State.

and briefly introduced in Chapter 1, subsection 1.7.2.4. The data used here consist of the number of incident leukemia cases in an eight-county region of central New York State, by census tract, for the period 1978–1982. There are 592 cases and 281 census tracts. Also provided are the x- and y-coordinates associated with the tract centroids and the 1980 census populations for each tract (in 1980, the total population of the eight-county region was 1,057,673). Centroids are shown in Figure 11.1; the cluster of centroids in the north is Syracuse; the cluster in the southern portion of the study area is Binghamton. The data are downloadable from www.wiley.com after searching for information on Waller and Gotway (2004).

Application of the previous methods yields Table 11.2. The units of σ are in terms of the length of the side of an average census tract (thus, $\sigma = 2$ implies that the smoothing bandwidth has a radius of approximately two census tracts). Interpretation of the table is as follows:

1. All of the global statistics U^2 are significant. A large degree of smoothing ($\sigma > 2.5$) begins to reduce the significance of the global statistic.
2. The maximum local statistic U_i is achieved with no spatial smoothing ($\sigma = 0$). In this nonspatial case, there is just one census

TABLE 11.2

Results of Statistical Tests for Data on Leukemia in Central New York State

σ	U^2	Max U_i	$U^*_{\alpha=.05}$	No. of Local Statistics $(U_i) > 1.96$	No. of Local Statistics $> U_{max,\alpha}$	Max U_i in Binghamton Cluster
0	430**	5.12	3.57	27	1	3.18
0.5	796**	5.10	3.54	69	7	4.03
0.75	945**	4.94	3.49	71	17	4.13
1	1028**	4.80	3.43	58	26	4.56
1.25	1066**	4.67	3.38	60	29	4.67
1.5	1076**	4.62	3.32	59	39	4.61
1.75	1071**	4.51	3.26	62	43	4.50
2	1060**	4.47	3.21	60	46	4.37
2.25	1047**	4.41	3.16	60	49	4.25
2.5	1036**	4.33	3.11	61	51	4.14
2.75	1026*	4.23	3.06	63	51	4.05
3	1019*	4.11	3.01	64	54	4.02

** $p < .01$.
* $p < .05$.

tract (near Cortland, just south of the center of the study area) that has a local statistic higher than the critical value of 3.57. This tract has 8.2 observed cases, and 1.6 cases are expected, based on population.

3. As smoothing is introduced, the maximum local statistic continues to exceed its critical value.* A new cluster emerges in and near Binghamton (see Figure 11.1, which shows subareas where local statistics exceed the critical value associated with the maximum local statistic). The cluster in the south (Binghamton) has 86.3 observed cases, and 55.1 are expected. The cluster to the north (near Cortland) consists of just two census tracts (of relatively large sizes); this cluster has 13.4 observed cases, and 3.7 are expected. The Binghamton cluster consists of 24 census tracts that, by themselves, do not display statistical significance; the cluster achieves a maximum local statistic when $\sigma = 1.25$ (shown in the last column of the table).

4. There are many local statistics that exceed the threshold of statistical significance ($U_{crit.} = 1.96$), where this threshold is appropriate if a single local test is being carried out. There are 27 such tracts in the unsmoothed, nonspatial case (where $281 \times 0.025 \approx 7$

---

* Note that the critical values obtained using Equations 11.16 and 11.17 may be somewhat inaccurate because the census tracts do not comprise a regular grid (see Section 5.10.5).

tracts would be expected to exceed this if there was no raised incidence). When smoothing is introduced, many more tracts would have significant local statistics. However, unless these local statistics are used in conjunction with prespecification of the tract of interest, it is more appropriate to use the maximum local statistics, which adjust for multiple hypothesis testing. A comparison of the 5th and 6th columns suggests that there are many tracts that when examined alone would be significant if they were specified as individual areas of interest, but for many of them, the magnitude of the significance is insufficient to lead to significance once the testing of multiple tracts is accounted for.

11.3 Associated Prospective Statistical Tests: Regional Surveillance for Quick Detection of Change

As we have seen in earlier chapters, cumulative sum methods allow for the quick detection of increases in a variable from one mean to another. They account for the multiple hypothesis testing that is associated with repeated assessment as new data become available.

In an aspatial context, we can consider monitoring either a single region, a set of regions, or a global statistic representing all of the regions. In a spatial context, we may wish to monitor either a single local statistic (comprising information on a particular region and surrounding regions), many local statistics simultaneously (by adjusting for the nonindependence of nearby local statistics), or a global statistic representing all of the regions. The nonspatial and spatial cases are treated in subsections 11.3.1 and 11.3.2, respectively.

11.3.1 Prospective Methods: Aspatial Case

For a single region, one can monitor the region-specific local statistic by maintaining the cumulative sum based on the z-scores (e.g., defined as in Equation 11.1 or 11.18):

$$S_{i,t} = \max(0, S_{i,t-1} + z_t - k) \tag{11.19}$$

where k is a parameter equal to one-half the size of the expected change (represented in standard deviations); this is usually set to 0.5, implying that the test aims at detecting a change of one standard deviation.

Then choose a value ARL_0 equal to the average number of observations between false alarms (similar to a Type I error rate). Based on this choice, derive a threshold

$$h = \frac{ARL_0 + 4}{ARL_0 + 2} \ln\left(\frac{ARL_0}{2} + 1\right) - 1.166 \qquad (11.20)$$

(This is appropriate when $k = \frac{1}{2}$; more generally, Equation 7.5 [Chapter 7] is used to find the threshold.) A significant increase is signaled when the cusum $S_{i,t}$ exceeds this threshold.

To monitor m regions simultaneously, separate cusum charts can be maintained for each region (using the z-scores for each region), and the threshold is based on $m \times ARL_0$ instead of ARL_0, so that false alarms across the regional system occur at the desired frequency. This is a Bonferroni adjustment and assumes independent charts.

As noted in Chapter 8, subsection 8.3.2, one could also monitor the global, chi-square goodness-of-fit statistic in an aspatial context, using cusum methods for chi-square random variables. In particular, the cusum (Chapter 8, Equations 8.4 and 8.5) can be used (together with the threshold to be obtained via Equation 8.6) to detect rapidly an upward shift in the global statistic.

Alternatively, if new data are in the form of the regional locations of each new case as it occurs, the approach outline in Chapter 10, subsection 10.4.1 (and based on Rogerson 1997), can be used, adopting the identity matrix for the weight matrix.

11.3.2 Prospective Methods: Spatial Case

The previous subsection outlined cusum approaches to surveillance when the spatial scale of interest was that of the regions for which data are collected. If monitoring on a large spatial scale is desired (e.g., to quickly detect clusters that may emerge on that larger scale), then spatial methods that use local statistics comprising neighborhood information are needed.

Monitoring of the local statistics based on neighborhoods, as defined by Equation 11.7 or Equation 11.8 (after the latter has been standardized by dividing by the square root of the variance given in Equation 11.4), is straightforward because these local statistics have standard normal distributions. In particular, Equation 11.19 and Equation 11.20 may be used.

If m separate charts for local statistics are maintained, the Bonferroni adjustment, made by using m^*ARL_0 to determine the threshold, is conservative (i.e., there will be fewer false alarms, and hence also fewer "real" detections than desired) because local statistics are correlated with one another. Instead, when Gaussian weights are used for the local statistics, we base the threshold on e^*ARL_0, where

$$e = \frac{m}{1+0.81\sigma^2} \qquad (11.21)$$

and where σ is the standard deviation of the Gaussian kernel. The standard deviation is expressed here as a unitless quantity and is interpreted in terms of the number of average spatial units associated with the width of the kernel. This quantity is found by dividing the unstandardized value of the standard deviation by the square root of the average area of a spatial unit (i.e., the original standard deviation used for the smoothing kernel is divided by $\sqrt{A/m}$, where A is the size of the study area).

Finally, to monitor the global spatial statistic (Equation 11.9 or Equation 11.10), one can make use of the fact that they have chi-square distributions, and use Equations 8.3–8.6, replacing $m - 1$ with the degrees of freedom (Equation 11.15) and χ_t^2 with U^2. As interpretation of the alternative hypothesis is not as straightforward as in the aspatial case (i.e., it is not as clear how to specify the value to which U^2 will change when the null hypothesis is no longer true), another approach is simply to use a cusum to monitor the z-scores associated with the normal approximation given in Equation 11.14.

References

Adler, R.J. 1981. *The Geometry of Random Fields*. New York: Wiley.

Aickin, M. and Gensler, H. 1996. Adjusting for multiple testing when reporting research results: the Bonferroni vs. Holm methods. *American Journal of Public Health*, 86: 726–728.

Alwan, L. 2000. Designing an effective exponential cusum chart without the use of nomographs. *Communications in Statistics—Theory and Methods*, 29: 2879–2893.

Alt, F.B. 1984. Multivariate quality control. In *The Encyclopedia of Statistical Sciences*. Eds., S. Kotz, N.L. Johnson, and C.R. Read. New York: John Wiley. pp. 110–112.

Anselin, L. 1995. Local indicators of spatial association—LISA. *Geographical Analysis*, 27: 93–115.

Anselin, L., Syabri, I., and Kho, Y. 2006. GeoDa: an introduction to spatial data analysis. *Geographical Analysis*, 38(1): 5–22.

Babcock, G., Talbot, T., Rogerson, P., and Forand, S. 2005. Use of CUSUM and Shewart charts to monitor regional trends of birth defects reports in New York State. *Birth Defects Research (Part A)*, 73: 668–677.

Bachi, R. 1963. Standard distance measures and related methods for spatial analysis. *Papers of the Regional Science Association*, 10: 83–132.

Bailey, A. and Gatrell, A. 1995. *Interactive Spatial Data Analysis*. Essex: Longman (published in the U.S. by Wiley).

Barbujani, G. 1987. A review of statistical methods for continuous monitoring of malformation frequencies. *European Journal of Epidemiology*, 3: 67–77.

Barton, D.E. and David, F.N. 1966. The random intersection of two graphs. In *Research Papers in Statistics: Festschrift for J. Neyman*. Ed., F.N. David. New York: Wiley. pp. 445–459.

Berry, K.J. and Mielke, P.W. 1998. A rapid recursion method for computing cumulative hypergeometric probability values. *Perceptual and Motor Skills*, 87: 51–55.

Besag, J. and Newell, J. 1991. The detection of clusters in rare diseases. *Journal of the Royal Statistical Society, Series A*, 154: 143–155.

BioMedware 2005. Homepage. BioMedware, Inc. Accessed on January 21, 2008. http://www.biomedware.com/index.html.

Bivand, R. 2006. Implementing spatial data analysis software tools in R. *Geographical Analysis*, 38(1): 23–40.

Blom, G. 1954. Transformations of the binomial, negative binomial, poisson, and χ^2 distributions. *Biometrika* 41: 302–316.

Bodiwala, D., Luscombe, C.J., Liu, S., Saxby, M., French, M., Jones, P.W., Fryer, A.A., and Strange, R.C. 2003. Prostate cancer risk and exposure to ultraviolet radiation: further support for the protective effect of sunlight. *Cancer Letters*, 192(2):145–149.

Bollobas, B. 1985. *Random Graphs*. New York: Academic Press.

Borror, C.M., Keats, J.B., and Montgomery, D.C. 2003. Robustness of the time between events CUSUM. *International Journal of Production Research*, 41: 3435–3444.

Bourke P.D. 2001. Sample size and the binomial CUSUM control chart: the case of 100% inspection. *Metrika* 53: 51–70.

Bowman, A.W. and Azzalini, A. 1997. *Applied Smoothing Techniques for Data Analysis: The Kernel Approach with S-Plus Illustrations*. Oxford: Clarendon Press.

Breslow, N. and Clayton, D.G. 1993. Approximate inference in generalized linear mixed models. *Journal of the American Statistical Association*, 88: 9–25.

Breslow, N.E. and Day, N.E. 1987. *Statistical Methods in Cancer Research. Volume II. The Design and Analysis of Cohort Studies*. Lyon: International Agency for Research on Cancer.

Brook, D. and Evans, D.A. 1972. An approach to the probability distribution of CUSUM run length. *Biometrika*, 59: 539–549.

Brown, B.W. and Russell, K. 1997. Methods correcting for multiple testing: operating characteristics. *Statistics in Medicine*, 16: 2511–2528.

Brunsdon, C. 1995. Estimating probability surfaces for geographical points data: an adaptive algorithm. *Computers and Geosciences*, 21: 877–894.

Bunk, S. 2002. Early warning: U.S. scientists counter bioterrorism with new electronic sentinel systems. *The Scientist*, 16(9): 14.

Castro, M.C. and Singer, B.H. 2006. Controlling the false discovery rate: a new application to account for multiple and independent tests in local statistics of spatial association. *Geographical Analysis*, 38: 180–208.

Centers for Disease Control and Prevention (CDC). 2002. Infant mortality and low birth weight among black and white infants—United States, 1980–2000. *Morbidity and Mortality Weekly Report*, 51(27): 589–592.

Chapeau-Blondeau, F. and Monir, A. 2002. Numerical evaluation of the Lambert W function and applications to generation of generalized Gaussian noise with exponent ½. *IEEE Transactions of Signal Processing*, 50: 2160–2165.

Chen, R. 1978. A surveillance system for congenital malformations. *Journal of the American Statistical Association*, 73: 323–327.

Chen, R., Mantel, N., Connelly, R.R., and Isacson, P. 1982. A monitoring system for chronic diseases. *Methods of Information in Medicine*, 21: 86–90.

Chen, R., Connelly, R.R., and Mantel, N. 1993. Analysing post alarm data in a monitoring system, in order to accept or reject the alarm. *Statistics in Medicine*, 12: 1807–1812.

Chen, R., Iscovich, J., and Goldbourt, U. 1997. Clustering of leukaemia cases in a city in Israel. *Statistics in Medicine*, 16: 1873–1887.

Choynowski, M. 1959. Maps based on probabilities. *Journal of the American Statistical Association*, 54: 385–388.

Clark, P.J. and Evans, F.C. 1954. Distance to nearest neighbor as a measure of spatial relationships in populations. *Ecology*, 35: 445–453.

Clarke, R.V.G. 1980. Situational crime prevention: theory and practice. *British Journal of Criminology*, 20: 136–147.

Clarke, R.V.G. 1992. *Situational Crime Prevention: Successful Case Studies*. Harrow and Heston, New York.

Cliff, A. and Ord, J.K. 1981. *Spatial Process: models and applications*, London: Pion Ltd.

Collica, R.S., Ramirez, J.G. and Taam, W. 1996. Process monitoring in integrated circuit fabrication using both yield and spatial statistics. *Quality and Reliability Engineering International*, 12: 195–202.

Corless, R.M., Jeffrey, D.J., and Knuth, D.E. 1997. A sequence of series for the Lambert W function. Proceedings ISAAC '97, Maui. Ed., W.W. Kuechlin. pp. 197–204.

Corless, R.M., Gonnet, G.H., Hare, D.E.G., Jeffrey, D.J., and Knuth, D.E. 1996. On the Lambert W function. *Advances in Computational Mathematics*, 5: 329–359.

Crawford, E.D. 2003. Epidemiology of prostate cancer. *Urology* 62: 3–12.

Cressie, N. and Read, T.R.C. 1984. Multinomial goodness-of-fit tests. *Journal of the Royal Statistical Society Series B* 46: 440–464.

Cressie, N. 1993. *Statistics for Spatial Data.* New York: Wiley.

Crosier, R.B. 1988. Multivariate generalizations of cumulative sum quality-control schemes. *Technometrics*, 30: 291–303.

Curtiss, J. and McIntosh, R. 1950. The interrelations of certain analytic and synthetic phytosociological characters. *Ecology*, 31: 434–455.

Cuzick, J. and Edwards, R. 1990. Spatial clustering for inhomogeneous populations (with discussion). *Journal of the Royal Statistical Society, Series B*, 52: 73–104.

Daniels, M.J. and Kass, R.E. 2001. Shrinkage estimators for covariance matrices. *Biometrics*, 57: 1173–1184.

David, H.A. 1956. On the application to statistics of an elementary theorem in probability. *Biometrika*, 43: 85–91.

Devesa, S.S., Grauman, D.J., Blot, W.J., Pennello, G., Hoover, R.N., and Fraumeni, J.F., Jr. 1999. *Atlas of Cancer Mortality in the United States, 1950–1994.* Washington, DC: U.S. Government Printing Office [NIH Publ No. (NIH) 99-4564].

Diggle, P.J. 1990. A point process modeling approach to raised incidence of a rare phenomenon in the vicinity of a prespecified point. *Journal of the Royal Statistical Society, Series A*, 153: 349–362.

Diggle, P.J. and Chetwynd, A.G. 1991. Second-order analysis of spatial clustering for inhomogeneous populations. *Biometrics* 47: 1155–1163.

Diggle, P.J. and Rowlingson, B.S. 1994. A conditional approach to point process modeling of elevated risk. *Journal of the Royal Statistical Society, Series A*, 157: 433–440.

Eck, J.E. 1995. Examining routine activity theory: a review of two books. *Justice Quarterly*, 12(4): 783–797.

Eilon, S., Watson-Gandy, C.D.T., and Christofides, N. 1971. *Distribution Management: Mathematical Modeling and Practical Analysis.* New York: Hafner.

Farrington, C.P. and Beale, A.D. 1998. The detection of outbreaks of infectious disease. In *GEOMED '97, International Workshop on Geomedical Systems.* Eds., L. Gierl, A.D. Cliff, A. Valleron, P. Farrington, and M. Bull. Stuttgart: B.G. Teubner. pp. 97–117.

Forsberg, L., Bonetti, M., Jeffery, C., Ozonoff, A., and Pagano, M. 2005. Distance-based methods for spatial and spatio-temporal surveillance. In *Spatial and Syndromic Surveillance for Public Health.* Eds., A.B. Lawson and K. Kleinman. Chichester, England: Wiley. pp. 133–152.

Fotheringham, A.S. and Wong, D.W.S. 1991. The modifiable areal unit problem in multivariate statistical analysis. *Environment and Planning A*, 23(7): 1025–1044.

Fotheringham, A.S. and Zhan, F.B. 1996. A comparison of three exploratory methods for cluster detection in spatial point patterns. *Geographical Analysis*, 28: 200–218.

Freeman, M.F. and Tukey, J.W. 1950. Transformations related to the angular and the square root. *Annals of Mathematical Statistics* 21: 607–611.

Frisén, M. and Sonesson, C. 2005. Optimal surveillance. In *Spatial and Syndromic Surveillance for Public Health*. Eds., A.B. Lawson and K. Kleinman. Chichester, England: Wiley. pp. 31–52.

Fuchs, C. and Kenett, R. 1980. A test for detecting outlying cells in the multinomial distribution and two-way contingency tables. *Journal of the American Statistical Association*, 75: 395–398.

Galambos, J. 1975. Methods for proving Bonferroni type inequalities. *Journal of the London Mathematical Society*, 9: 561–564.

Gan, F.F. 1994. Design of optimal exponential CUSUM control charts. *Journal of Quality Technology*, 26: 109–124.

Geary, R.C. 1954. The contiguity ratio and statistical mapping. *The Incorporated Statistician*, 5(3): 115–145.

Getis, A. and Ord, J. 1992. The analysis of spatial association by use of distance statistics. *Geographical Analysis*, 24: 189–206.

Greig-Smith, P. 1964. *Quantitative Plant Ecology*. London: Butterworth and Company.

Green, L. 1995. Cleaning up drug hot spots in Oakland, California: the displacement and diffusion effects. *Justice Quarterly*, 12(4): 737–754.

Gregorio, D.I., DeChello, L.M., Samociuk, H., and Kulldorff, M. 2005. Lumping or splitting: seeking the preferred areal unit for health geography studies. *International Journal of Health Geographics* 4/1/6. Accessible at http://www.ij-healthgeographics.com/content/4/1/6.

Griffith, D.A. 1987. *Spatial autocorrelation: a primer*. Washington, DC: Association of American Geographers.

Griffith, D.A. 2006. Hidden negative spatial autocorrelation. *Journal of Geographical Systems*, 8: 335–355.

Gumbel, E.J. 1958. *Statistics of Extremes*. New York: Columbia University Press.

Haining, R. 1990. *Spatial Data Analysis in the Social and Environmental Sciences*. Cambridge: Cambridge University Press.

Haining, R. 2003. *Spatial Data Analysis: Theory and Practice*. Cambridge: Cambridge University Press.

Han, D., Rogerson, P.A., Bonner, M.R., Nie, J., Vena, J.E., Muti, P., Trevisan, M., and Freudenheim, J.L. 2005. Assessing spatio-temporal variability of risk surfaces using residential history data in a case control study of breast cancer. *International Journal of Health Geographics* 4/1/9. Accessible at www.ij-healthgeographics.com/content/4/1/9.

Han, D. and Rogerson, P. 2003. Application of a GIS-based statistical method to assess spatio-temporal changes in breast cancer clustering in the northeastern United States. In *Geographic Information Systems and Health Application*. Eds., O. Khan and R. Skinner. pp. 114–138.

Hansen, M.H., Nair, V.N., and Friedman, D.J. 1997. Monitoring wafer map data from integrated circuit fabrication processes for spatially clustered defects. *Journal of the American Statistical Association*, 39: 241–253.

Hawkins, D.M. and Olwell, D.H. 1998. *Cumulative Sum Charts and Charting for Quality Improvement*. New York: Springer-Verlag.

Haybittle, J., Yuen, P., and Machin, D. 1995. Multiple comparisons in disease mapping. *Statistics in Medicine*, 14: 2503–2505.

Healy, J.D. 1987. A note on multivariate CUSUM procedures. *Technometrics*, 29: 409–412.

Hill, G.B., Spicer, C.C., and Weatherall, J.A.C. 1968. The computer surveillance of congenital malformations. *British Medical Journal*, 24: 215–218.

Hirotsu, C. 1993. Beyond analysis of variance techniques: some applications in clinical trials. *International Statistical Review*, 61: 183–201.

Hochberg, Y. 1988. A sharper Bonferroni procedure for multiple tests of significance. *Biometrika*, 75: 800–803.

Hochberg, Y. and Benjamini, Y. 1990. More powerful procedures for multiple significance testing. *Statistics in Medicine*, 9: 811–818.

Holm, S. 1979. A simple sequentially rejective multiple test procedure. *Scandinavian Journal of Statistics*, 6: 65–70.

Hunter, J.S. 1986. The exponentially weighted moving average. *Journal of Quality Technology*, 18: 203–210.

Hutwagner, L.C., Maloney, E.K., Bean, N.H., Slutsker, L., and Martin, S.M. 1997. Using laboratory-based surveillance data for prevention: an algorithm for detecting *Salmonella* outbreaks. *Emerging Infectious Diseases*, 3: 395–400.

Jackson, J.E. 1985. Multivariate quality control. *Communications in Statistics—Theory and Methods*, 14: 2657–2688.

Jemal, A., Kulldorff, M., Devesa, S.S., Hayes, R.B., and Fraumeni, J.F. 2002. A geographic analysis of prostate cancer mortality in the United States, 1970–89. *International Journal of Cancer*. 101: 168–174.

Kenney, L.W. and Forde, D.R. 1990. Routine activities and crime: an analysis of victimization in Canada. *Criminology*, 28: 137–151.

Kelsall, J. and Diggle, P. 1995. Non-parametric estimation of spatial variation in relative risk. *Statistics in Medicine*, 14: 2335–2342.

Knox, G. 1964. The detection of space–time interactions. *Applied Statistics*, 13: 25–29.

Koehler, K.J. and Larntz, K. 1980. An empirical investigation of goodness-of-fit statistics for sparse multinomials. *Journal of the American Statistical Association*, 75: 336–344.

Kleinman, K., Lazarus, R., and Platt, R. 2004. A generalized linear mixed models approach for detecting incident clusters of disease: biological terrorism and other surveillance. *Epidemiology* 2004; *American Journal of Epidemiology*, 156: 217–224.

Kulldorff, M. 1998. Statistical methods for spatial epidemiology: tests for randomness. In *GIS and Health*. Eds., A.C. Gatrell and M. Loytonen. London: Taylor & Francis p. 49–62.

Kulldorff, M. 1999. Spatial scan statistics: models, calculations, and applications, In *Scan Statistics and Applications*. Eds., J. Glaz and N. Balakrishnan. Boston: Birkhauser. pp. 303–322.

Kulldorff, M. 2001. Prospective time periodic geographical disease surveillance using a scan statistic. *Journal of the Royal Statistical Society, Series A*, 164: 61–72.

Kulldorff, M. 2006. *SaTScan™ User Guide for version 7.0*. SaTScan™. Accessed on August 13, 2007. Available at: http://www.satscan.org/.

Kulldorff, M. and Hjalmers, U. 1999. The Knox method and other tests for space-time interactions. *Biometrics*, 55: 544–552.

Kulldorff, M., Huang, L., Pickle, L., and Duczmal, L. 2006. An elliptic spatial scan statistic. *Statistics in Medicine*, 25(22): 3929–3943.

Kulldorff, M. and Nagarwalla, N. 1995. Spatial disease clusters: detection and inference. *Statistics in Medicine*, 14(8): 799–810.

Kulldorff, M., Feuer, E.J., Miller, B.A., and Freedman, L.S. 1997. Breast cancer clusters in the United States: a geographic analysis. *American Journal of Epidemiology*, 146: 161–170.

Lagazio, C., Marchi, M., and Biggeri, A. 1996. The association between risk of disease and point sources of pollution: a test for case-control data. *Statistica Applicata*, 8: 343–356.

Lawson, A.B. 1993. On the analysis of mortality events associated with a pre-specified fixed point. *Journal of the Royal Statistical Society, Series A*, 156: 363–377.

Lawson, A.B. 2001. *Statistical Methods in Spatial Epidemiology*. New York: Wiley.

Lawson, A.B. 2005. Advanced modeling for surveillance: clustering of relative risk changes. In *Spatial and Syndromic Surveillance for Public Health*. Eds., A.B. Lawson and K. Kleinman. Chichester, England: Wiley. pp. 223–243.

Lee, G. and P. Rogerson. 2007. Monitoring global spatial statistics. *Stochastic Environmental Research and Risk Assessment*, 21 (5): 545–553.

Le Strat, Y. 2005. Overview of temporal surveillance. In *Spatial and Syndromic Surveillance for Public Health*. Eds., A.B. Lawson and K. Kleinman. Chichester, England: Wiley. pp. 13–29.

Levine, N. 2007. *CrimeStat: A Spatial Statistics Program for the Analysis of Crime Incident Locations, version 3.1*. Ned Levine & Associates, Houston, TX, and the National Institute of Justice, Washington DC. Accessed on January 21, 2008. http://www.icpsr.umich.edu/CRIMESTAT/.

Lowry, C.A. and Montgomery, D.C. 1995. A review of multivariate control charts. *IIE Transactions*, 27: 800–810.

Lucas, J.M. and Crosier, R.B. 1982. Fast initial response for CUSUM quality control schemes: give your CUSUM a head start. *Technometrics*, 24: 199–205.

Lucas, J.M. 1985. Counted data cusums. *Technometrics*, 27: 129–144.

Lucas, J.M. and Saccucci, M.S. 1990. Exponentially weighted moving average control schemes: properties and enhancements. *Technometrics*, 32: 1–12.

Mandl, K.D., Overhage, J.M., Wagner, M.M., Lober, W.B., Sebastiani, P., Mostashari, F., Pavlin, J.A., Gesteland, P.H., Treadwell, T., Koski, E., Hutwagner, L., Buckeridge, D.L., Aller, R.D., and Grannis, S. 2003. Implementing syndromic surveillance: A practical guide informed by the early experience. *Journal of the American Medical Informatics Association* (November 21, 2003). Available at: http://www.jamia.org/cgi/reprint/M1356v1.pdf.

Mardia, K.V. 1972. *Statistics of Directional Data*. London: Academic Press.

McGrew, J.C. and Monroe, C.B. 1993. *An Introduction to Statistical Problem Solving in Geography*. Dubuque, IA: William C. Brown Publishers.

Mercer, W.B. and Hall, A.D. 1911. The experimental error of field trials. *Journal of Agricultural Science*, 4: 107–132.

Montgomery, D. (1996). *Introduction to Statistical Quality Control*. New York: John Wiley.

Moran, P.A.P. 1948. The interpretation of statistical maps. *Journal of the Royal Statistical Society, Series B*, 10: 245–251.

Mostashari, F. and Hartman, J. 2003. Syndromic surveillance: a local perspective. *Journal of Urban Health*, 80(1): i1–i7.

National Center for Health Statistics. Compressed Mortality File, 1968–1998.

National Syndromic Surveillance Conference, September, 2002. Links to presentations and posters at http://www.nyam.org/events/syndromicconference/

agenda.shtml and http://www.nyam.org/events/syndromicconference/poster.shtml.

Oden, N. 1995. Adjusting Moran's *I* for population density. *Statistics in Medicine*, 14: 17–26.

Openshaw, S. 1984. *The Modifiable Areal Unit Problem*. Nowich: Geo Books United Kingdom.

Openshaw, S., Charlton, M., Wymer, C., and Craft, A. 1987. A mark 1 geographical analysis machine for the automated analysis of point data sets. *International Journal of Geographical Information Systems*, 1: 335–358.

Ord, J. and Getis, A. 1995. Local spatial autocorrelation statistics: distributional issues and an application. *Geographical Analysis*, 27: 286–306.

Page, E.S. 1954. Continuous inspection schemes. *Biometrika*, 41: 100–115.

Pignatiello, J.J. Jr. and Runger, GC. 1990. Comparisons of multivariate CUSUM charts. *Journal of Quality Technology*, 22: 173–186.

Qiu, M. and Hawkins, D. 2003. A nonparametric multivariate CUSUM procedure for detecting shifts in all directions. *Journal of the Royal Statistical Society Series D*, 52: 151–164.

Raubertas, R.F. 1989. An analysis of disease surveillance data that uses the geographic locations of the reporting units. *Statistics in Medicine*, 8: 267–271.

Roberts, S.W. 1959. Control chart tests based on geometric moving averages. *Technometrics*, 1(3): 239–250.

Roberts, S.W. 1966. A comparison of some control chart procedures. *Technometrics*, 8: 411–430.

Rogerson, P. 1997. Surveillance methods for monitoring the development of spatial patterns. *Statistics in Medicine*, 16: 2081–2093.

Rogerson, P. 1999. The detection of clusters using a spatial version of the chi-squared goodness-of-fit test. *Geographical Analysis*, 31: 130–147.

Rogerson, P. 2000. The monitoring of point incidents around a fixed location. In *Geography and Medicine: GEOMED '99*. Eds., A. Flahaut, L. Toubiana, and A.J. Valleron. Paris: Elsevier. pp. 18–27.

Rogerson, P. 2001a. A statistical method for the detection of geographic clustering. *Geographical Analysis*, 33: 215–227.

Rogerson, P. 2001b. Monitoring point patterns for the development of space-time clusters. *Journal of the Royal Statistical Society, Series A*, 164: 87–96.

Rogerson, P. 2004a. The application of new spatial statistical methods to the detection of geographical patterns of crime. *Applied GIS and Spatial Analysis*. Eds., J. Stillwell and G. Clarke. Chichester: John Wiley & Sons. pp. 151–168.

Rogerson, P. 2004b. The statistical significance of the maximum local statistic. In *Spatial Econometrics and Spatial Statistics*. Eds., A. Getis, J. Mur, and H. Zoller. New York: Palgrave Macmillan Press. pp. 250–264.

Rogerson, P. 2005a. A set of associated statistical tests for the detection of spatial clustering. *Ecological and Environmental Statistics*, 12(3): 275–288.

Rogerson, P. 2005b. Monitoring spatial maxima. *Journal of Geographical Systems*, 7: 101–114.

Rogerson, P. 2005c. Spatial surveillance and cumulative sum methods. *Spatial and Syndromic Surveillance for Public Health*. Eds. A.B. Lawson and K. Kleinman. Chichester, England: John Wiley and Sons. pp 95–114.

Rogerson, P. 2006a. Statistical methods for the detection of spatial clustering in case-control data. *Statistics in Medicine*, 25: 811–823.

Rogerson, P. 2006b. Formulas for the design of CUSUM quality control charts. *Communications in Statistics—Theory and Methods*, 35: 373–383.

Rogerson, P. 2006c. *Statistical Methods for Geography*, 2nd edition. London: Sage Publications.

Rogerson, P. 2009a. Monitoring changes in spatial patterns. *Handbook of Spatial Analysis*. Eds., A.S. Fotheringham and P. Rogerson, London: Sage Publications. pp. 310–320.

Rogerson, P. 2009b. Health surveillance around prespecified locations using case-control data. *Essays in Honor of Arthur Getis*. Eds., S. Rey and L. Anselin. Springer. Forthcoming.

Rogerson, P., Sinha, G., and Han, D. 2006. Recent changes in the spatial pattern of prostate cancer in the United States. *American Journal of Preventive Medicine*, 30: S50–S59.

Rogerson, P. and Sun, Y. 2001. Spatial monitoring of geographic patterns: an application to crime analysis. *Computers, Environment, and Urban Systems*, 25/6: 539–556.

Rogerson, P. and Yamada, I. 2004a. Monitoring change in spatial patterns of disease: comparing univariate and multivariate cumulative sum approaches. *Statistics in Medicine*, 23: 2195–2214.

Rogerson, P. and Yamada, I. 2004b. Approaches to syndromic surveillance when data consist of small regional counts. *Morbidity and Mortality Weekly Report*, 53(Supplement): 79–85.

Rosenfeld, A. and Kak, A.C. 1982. *Digital Picture Processing, Volume 2*. Orlando: Academic Press.

Rossi, G., Lampugnani, L., and Marchi, M. 1999. An approximate CUSUM procedure for surveillance of health events. *Statistics in Medicine*, 18: 2111–2122.

Rushton, G. and Lolonis, P. 1996. Exploratory spatial analysis of birth defect rates in an urban population. *Statistics in Medicine*, 15: 717–726.

Ryan, T.P. 2000. *Statistical Methods for Quality Improvement*. New York: Wiley.

Sasvari, Z. 1999. Tight bounds for the normal distribution. *American Mathematical Monthly*, 106: 76.

Schweder, T. and Spjotvoll, E. 1982. Plots of *p*-values to evaluate many tests simultaneously. *Biometrika*, 69: 493–502.

Scott, D.W. 1992. *Multivariate Density Estimation: Theory, Practice, and Visualization*. New York: Wiley.

Sherman, L.W. 1995. Hot spots of crime and criminal careers of places. In *Crime and Place*. Eds., J.E. Eck and D. Weisburd. Monsey, New York: Criminal Justice Press. pp. 35–52.

Shiryaev, A.N. 1963. On optimum methods in quickest detection problems. *Theory of Probability and its Applications*, 8: 22–46.

Sidak, A. 1968. On multivariate normal probabilities of rectangles: their dependence on correlations. *Annals of Mathematical Statistics*, 39: 1425–1434.

Siegmund, D.O. 1985. *Sequential Analysis: Test and Confidence Intervals*. New York: Springer-Verlag.

Siegmund, D.O. and Worsley, K.J. 1995. Testing for a signal with unknown location and scale in a stationary Gaussian random field. *Annals of Statistics*, 23: 608–639.

Skellam, J.G. 1952. Studies in statistical ecology. I. Spatial pattern. *Biometrika*, 39: 346–362.

Sonesson, C. and Bock, D. 2003. A review and discussion of prospective statistical surveillance in public health. *Journal of the Royal Statistical Society, Series A*, 166: 5–21.

Srivastava, M.S. 1997. CUSUM procedure for monitoring variability. *Communications in Statistics: Theory and Methods*, 12: 2905–2926.

Srivastava, M.S. and Worsley, K.J. 1986. Likelihood ratio tests for a change in the multivariate normal mean. *Journal of the American Statistical Association*, 81: 199–204.

Spatial and Temporal Analysis of Crime (STAC). Illinois Criminal Justice Information Authority.

Stern, H. and Cressie, N. 1999. Inference for extremes in disease mapping. In *Disease Mapping and Risk Assessment for Public Health*. Eds., A. Lawson, A. Biggeri, D. Böhning, E. Lesaffre, J.F. Viel, and R. Bertollini. New York: Wiley. pp. 63–84.

Stone, R. 1988. Investigation of excess environmental risks around putative sources: statistical problems and a proposed test. *Statistics in Medicine*, 7: 649–660.

Student 1907. On the error of counting with a haemacytometer. *Biometrika*, 5: 351–360.

Tango, T. 1995. A class of tests for detecting "general" and "focused" clustering of rare diseases. *Statistics in Medicine*, 7: 649–660.

Tango T. 1999. Comparison of general tests for disease clustering. In *Disease Mapping and Risk Assessment for Public Health*. Eds., A. Lawson et al. New York: John Wiley. pp. 111–117.

Tango, T. 2000. A test for spatial disease clustering adjusted for multiple testing. *Statistics in Medicine*, 19: 191–204.

Taylor, R. 1997. Crime and small-scale places: what we know, what we can prevent, and what else we need to know. In NIJ Research Forum, *Crime and Place: Plenary Papers of the 1997 Conference on Criminal Justice Research and Evaluation*. National Institute of Justice, Washington, DC.

Turnbull, B.W., Iwano, E.J., Burnett, W.S., Howe, H.L., and Clark, L.C. 1990. Monitoring for clusters of disease: application to leukemia incidence in upstate New York. *American Journal of Epidemiology*, 132: S136–143.

U.S. Census Bureau 2001. Centers of population computation: 1950, 1960, 1970, 1980, 1990, and 2000. Available at http://www.census.gov/geo/www centers_pop.pdf. Accessed on 03/13/2008.

Venables, W.N., Smith, D.M., and the R Development Core Team. 2007. An introduction to R—notes on R: a programming environment for data analysis and graphics. The R Project for Statistical Computing. Accessed on January 21, 2008. Available at: http://www.r-project.org/.

Waller, L. and Lawson, A. 1995. The power of focused tests to detect disease clustering. *Statistics in Medicine*, 14: 2291–2308.

Waller, L.A., Turnbull, B.W., Clark, L.C., and Nasca, P. 1992. Chronic disease surveillance and testing of clustering of disease and exposure: application to leukemia incidence and TCE-contaminated dumpsites in upstate New York. *Environmetrics*, 3: 281–300.

Waller, L.A., Turnbull, B.W., Clark, L.C., and Nasca, P. 1994. Spatial pattern analyses to detect rare disease clusters. In *Case Studies in Biometry*. Eds., N. Lange, L. Ryan, L. Billard, D. Brillinger, L. Conquest, and J. Greenhouse. New York: Wiley. pp. 3–23.

Waller, L. and Gotway, C. 2004. *Applied Spatial Statistics for Public Health Data*. New York: Wiley.

Walsh, T. 2002. Will health databases spot bioterror attacks? *Government Computer News*, 21(4), February 18, 2002. www.gcn.com.

Weatherall, J.A.C. and Haskey, J.C. 1976. Surveillance of malformations. *British Medical Journal*, 32: 39–44.

Weisburd, D. and Green, L. 1995. Policing drug hot-spots: the Jersey City drug market analysis experiment. *Justice Quarterly*, 12(4): 711–735.

Westfall, P.H. and Young, S.S. 1993. *Resampling-Based Multiple Testing: Examples and Methods for p-Value Adjustment*. New York: Wiley.

Wetherill, G.W. and Brown, D.W. 1991. *Statistical Process Control: Theory and Practice*. New York: Chapman & Hall.

White, C.H. and Keats, J.B. 1996. ARLs and higher order run length moments for Poisson CUSUM. *Journal of Quality Technology*, 28: 363–369.

Williams, E.H., Smith, P.G., Day, N.E., Geser, A., Ellice, J., and Tukei, P. 1978. Space-time clustering of Burkitt's lymphoma in the West Nile District of Uganda: 1961–1975. *British Journal of Cancer*, 37: 109–122.

Wolter, C. 1987. Monitoring intervals between rare events: a cumulative score procedure compared with Rina Chen's sets technique. *Methods of Information in Medicine*, 26: 215–219.

Woodall, W.H. and Ncube, M.M. 1985. Multivariate CUSUM quality-control procedures. *Technometrics*, 27: 285–292.

Woodall, W.H. and Adams, B.M. 1993. The statistical design of CUSUM charts. *Quality Engineering*, 5: 559–570.

Worsley, K.J. 1979. On the likelihood ratio test for a shift in location of normal populations. *Journal of the American Statistical Association*, 74(366): 365–367.

Worsley, K.J. 1994. Local maxima and the expected Euler characteristic of excursion sets of χ^2, F and t fields. *Advances in Applied Probability*, 26: 13–42.

Worsley, K.J. 1996. The geometry of random images. *Chance*, 9(1): 27–40.

Worsley, K.J., Evans, A.C., Marrett, S., and Neelin, P. 1992. A three-dimensional statistical analysis for CBF activation studies in human brain. *Journal of Cerebral Blood Flow and Metabolism*, 12: 900–918.

Worsley, K.J., Andermann, M., Koulis, T., MacDonald, D., and Evans, A.C. 1999. Detecting changes in non-isotropic images. *Human Brain Mapping* 8: 98–101.

Author Index

A

Adams, B.M., 176–78
Adler, R.J., 125
Aickin, M., 122
Aller, R.D., 158
Alt, F.B., 284
Alwan, L., 172, 173, 176–78
Andermann, M., 133
Anselin, L., 12, 13, 85, 86, 94, 197, 207, 289
Azzalini, A., 38, 124

B

Babcock, G., 157
Bachi, R., 24
Bailey, A., 32, 34, 53, 204
Barbujani, G., 157
Barton, D. E., 136, 137
Beale, A.D., 157, 183, 185, 189
Bean, N.H., 167
Benjamini, Y., 123
Berry, K.J., 208
Besag, J., 11, 103, 107, 109–11, 187, 188, 243, 289
Biggeri, A., 98
Bivand, R., 13
Blom, G., 166
Blot, W.J., 9, 146
Bock, D., 157, 171
Bodiwala, D., 146
Bollobas, B., 262
Bonetti, M., 188
Bonner, M.R., 38, 39
Borror, C.M., 172
Bourke, P.D., 166
Bowman, A.W., 38, 124
Breslow, N. E., 187, 294
Brook, D., 232
Brown, B.W., 123
Brown, D.W., 160
Brunsdon, C., 219
Buckeridge, D.L., 158
Bunk, S., 186
Burnett, W.S., 111

C

Castro, M.C., 249
Chapeau-Blondeau, F., 177
Charlton, M., 108, 109
Chen R., 171, 185, 269
Chetwynd, A.G., 82
Choynowski, M., 6
Clark, L.C., 16, 90, 111, 297
Clark, P.J., 44
Clarke, R. V. G., 218
Clayton, D.G., 187
Cliff, A., 64
Collica, R., 185
Connelly, R.R., 171, 269
Corless R.M., 177, 241
Craft, A., 108, 109
Crawford, E.D., 145
Cressie, N., 6, 16, 32, 54, 125, 131, 260
Crosier, R.B., 174, 270, 271
Curtiss, J., 53
Cuzick, J., 16, 75, 78, 83

D

Daniels, M.J., 285, 287
David, F. N., 136, 137
David, H.A., 120, 136, 137
Day, N.E., 203, 204, 206, 212, 294
DeChello, L.M., 54
Devesa, S.S., 9, 146
Diggle, P.J., 37, 82, 96, 124, 238, 240
Duczmal, L., 116

E

Eck, J E., 218
Edwards, R., 16, 75, 78, 83
Eilon, S., 26
Ellice, J., 203, 204, 206, 212
Evans, A.C., 128
Evans, D.A., 232
Evans, F.C., 44

Subject Index

Printed and bound by CPI Group (UK) Ltd, Croydon, CR0 4YY
24/10/2024
01778278-0012